CLARKE, JOHN L
DESIGN OF CONCRETE ST[
00045

KV-429-849

624.012 C59

Proceedings of a seminar held at the Building Research Establishment, Garston, Watford, UK, on 29 and 30 November 1984, organised by the Institution of Structural Engineers Informal Study Group—Model Analysis as a Design Tool.

SCIENTIFIC COMMITTEE

Dr G. Somerville (*Chairman*)	*Cement and Concrete Association, UK*
Dr F. K. Garas (*Study Group Convener*)	*Taylor Woodrow Construction Ltd, UK*
Mr G. S. T. Armer	*Building Research Establishment, UK*
Dr J. L. Clarke	*Cement and Concrete Association, UK*
Dr W. G. Corley	*Portland Cement Association, USA*
Professor Dr-Ing. J. Eibl	*Karlsruhe University, FRG*
Professor J. E. Gibson	*City University, London, UK*
Professor H. G. Harris	*Drexel University, Philadelphia, USA*
Professor M. S. Mirza	*McGill University, Montreal, Canada*
Professor G. Oberti	*ISMES, Bergamo, Italy*

ORGANISING COMMITTEE

Dr F. K. Garas (*Chairman*)	*Taylor Woodrow Construction Ltd*
Mr G. S. T. Armer	*Building Research Establishment*
Dr J. L. Clarke	*Cement and Concrete Association*
Mr R. J. W. Milne	*Institution of Structural Engineers*
Mrs P. M. Rowley	*Building Research Establishment*
Mrs H. M. Stevenson	

DESIGN OF CONCRETE STRUCTURES

THE USE OF MODEL ANALYSIS

DESIGN OF CONCRETE STRUCTURES
THE USE OF MODEL ANALYSIS

Edited by

J. L. CLARKE
Cement and Concrete Association, UK

F. K. GARAS
Taylor Woodrow Construction Ltd, UK

and

G. S. T. ARMER
Building Research Establishment, UK

ELSEVIER APPLIED SCIENCE PUBLISHERS
LONDON and NEW YORK

ELSEVIER APPLIED SCIENCE PUBLISHERS LTD
Crown House, Linton Road, Barking, Essex IG11 8JU, England

Sole Distributor in the USA and Canada
ELSEVIER SCIENCE PUBLISHING CO., INC.
52 Vanderbilt Avenue, New York, NY 10017, USA

WITH 36 TABLES AND 231 ILLUSTRATIONS

© THE INFORMAL STUDY GROUP—MODEL ANALYSIS AS A DESIGN TOOL,
INSTITUTION OF STRUCTURAL ENGINEERS, 1985

British Library Cataloguing in Publication Data

Design of concrete structures: the use of model analysis.
1. Concrete construction   2. Architectural models
I. Clarke, J. L.   II. Garas, F. K.   III. Armer, G. S. T.
624.1'834'0228   TA681.5

Library of Congress Cataloging-in-Publication Data

Design of concrete structures.

Based on the proceedings of an international seminar organised by the Institution of Structural Engineers, Informal Study Group for "Model Analysis as a Design Tool" held in November 1984.
   Bibliography: p.
   Includes index.
   1. Concrete construction—Models—Testing—Congresses.   2. Structural design—Congresses.
I. Clarke, J. L.   II. Garas, F. K.   III. Armer, G. S. T.
IV. Institution of Structural Engineers (Great Britain). Informal Study Group for "Model Analysis as a Design Tool".
TA681.5.D44   1985   624.1'834   85-15865

ISBN 0-85334-387-X

The selection and presentation of material and the opinions expressed in this publication are the sole responsibility of the authors concerned.

All rights reserved. No part of this publication may be reproduced, stored in a retrieval system, or transmitted in any form or by any means, electronic, mechanical, photocopying, recording, or otherwise, without the prior written permission of the publisher.

Printed in Great Britain by Galliard (Printers) Ltd, Great Yarmouth

Preface

Physical models are widely used in the design of unconventional structures, such as nuclear containment vessels and offshore oil rigs, and also conventional structures subjected to complex loadings, such as multi-storey buildings under the effects of earthquakes. They are an invaluable complement to mathematical models. However, the use of physical models in the design process poses many problems in construction, testing and the subsequent interpretation of the results.

This book is based on the proceedings of an international seminar organised by the Institution of Structural Engineers Informal Study Group for 'Model Analysis as a Design Tool'. The seminar took place at the UK Building Research Establishment in November 1984 and attracted delegates from 8 different countries. The object of the seminar was to provide an international forum for the exchange of knowledge between research and design engineers on the use of models in the design of concrete structures.

Reproduced in this volume are 33 papers covering recent developments in modelling techniques, the reliability and accuracy of structural models, the analysis and interpretation of model test results and the application of model results in design. The discussion at the meeting is also reproduced and altogether the volume represents an international state-of-the-art report for the modelling of concrete structures.

J. L. CLARKE
F. K. GARAS
G. S. T. ARMER

Contents

Preface v

Section 1
RECENT DEVELOPMENTS IN MODELLING TECHNIQUES

1. Microconcrete for Structural Model Analysis . . . 1
   R. K. MÜLLER (*University of Stuttgart, FRG*)

2. Using Models for Studying Reinforced Fiber Concrete
   Behaviour 12
   R. J. CRAIG, O. LOY and C. RYDEN (*New Jersey Institute of Technology, USA*)

3. Development of a Small Aggregate Concrete for Structural
   Similitude of Slab–Column Connections 25
   R. N. SWAMY and F. M. FALIH (*University of Sheffield, UK*)

4. The Use of Pushout Tests to Simulate Shear Stud Connectors
   in Composite Beam Construction 38
   H. D. WRIGHT, P. W. HARDING and H. R. EVANS (*University College, Cardiff, UK*)

5. Small Scale Modelling Techniques for Investigating the
   Failure of a Bin Structure 48
   R. N. WHITE (*Cornell University, USA*)

6. Transient Loading of Structural Concrete 59
   H. K. CHEONG, A. H. AL-SHAIKH, N. N. AMBRASEYS and
   S. H. PERRY (*Imperial College of Science and Technology, London, UK*)

7. Production and Use of High Yield Model Reinforcement 69
   F. A. NOOR, S. RAVEENDRAN and W. EVANS (*North East London Polytechnic, UK*)

8. Adaptive Control Technique for the Simulation of Seismic Actions 79
   J. M. JERVIS PEREIRA and F. J. CARVALHAL (*Laboratório Nacional de Engenharia Civil, Portugal*)

   Commentary 88
   W. G. CORLEY (*Construction Technology Laboratories, Illinois, USA*)

Section 2
RELIABILITY AND ACCURACY OF STRUCTURAL MODELS

9. Comparison Between Reinforced Concrete Prototypes and Micro Concrete Models 92
   J. E. GIBSON (*City University, London, UK*)

10. An Evaluation of the Factors Contributing to Size Effects in Concrete 105
    P. CHANA (*Cement and Concrete Association, UK*)

11. Size Effects in Microconcrete Beams and Control Specimens . . 117
    S. MAJLESSI, F. A. NOOR and J. B. NEWMAN (*North East London Polytechnic, UK*)

12. Comparisons of Tests of Medium-scale Wall Assemblies and a Full-scale Building 129
    B. J. MORGAN, W. G. CORLEY (*Construction Technology Laboratories, Illinois, USA*) and H. HIRAISHI (*Building Research Institute, Japan*)

13. Dynamic Response of Models and Concrete Frame Structures . . 138
    D. P. ABRAMS (*University of Colorado, USA*)

14. Impact Testing on Microconcrete Models 149
    W. THOMA and R. K. MÜLLER (*University of Stuttgart, FRG*)

15. Impact Resistance of Reinforced Concrete Slabs . . 158
    A. J. WATSON and T. K. AL-AZAWI (*University of Sheffield, UK*)

16. Experimental and Numerical Investigations of Reinforced Concrete Structural Members Subjected to Impact Load . 168
    J. HERTER, E. LIMBERGER and K. BRANDES (*Bundesanstalt für Materialprüfung (BAM), Berlin, FRG*)

    Commentary 179
    G. SOMERVILLE (*Cement and Concrete Association, UK*)

Section 3
APPLICATION OF MODEL RESULTS IN DESIGN

17. Modeling Punching and Torsional Shear at Penetrations in R/C Slabs 181
    W. KIM and R. N. WHITE (*Cornell University, USA*)

18. Physical and Analytical Models for Flat Slab/Edge Column Connections 192
    D. J. CLELAND, S. G. GILBERT and A. E. LONG (*Queen's University, Belfast, UK*)

19. Experimental Investigations on the Problem of Lateral Buckling of Reinforced Concrete Beams 208
    H. TWELMEIER and D. BRANDMANN (*Technical University of Braunschweig, FRG*)

20. Experimental Behavior of Thick Pile Caps 220
    R. JIMENEZ-PEREZ (*University of Puerto Rico*), G. SABNIS (*Howard University, USA*) and A. B. GOGATE (*Ohio State University, USA*)

21. Model Analysis in Impact Research 230
    J. EIBL and M. FEYERABEND (*University of Karlsruhe, FRG*)

22. Modelling of Reinforced Concrete Containment Structures 247
    P. D. MONCARZ, J. D. OSTERAAS and A. M. CURZON (*Failure Analysis Associates, USA*)

23. Design of Concrete Slab Bridges against Flexural-shear Failure 257
R. J. COPE (*University of Liverpool, UK*)

Commentary 267
J. B. MENZIES (*Building Research Establishment, UK*)

Section 4
ANALYSIS AND INTERPRETATION OF MODEL TEST RESULTS

24. Laboratory Testing of Bored and Cast-in-situ Microconcrete Piles in Clay to Study Shaft Adhesion 268
W. F. ANDERSON, J. A. SULAIMAN (*University of Sheffield, UK*) and K. Y. YONG (*National University of Singapore*)

25. Model Analysis of Grouted Connections for Use in Construction and Repair of Offshore Structures . . . 277
L. F. BOSWELL and C. D'MELLO (*City University, London, UK*)

26. Reinforced Concrete Arch- and Box-type Structures under Severe Dynamic Loads 290
T. KRAUTHAMMER (*University of Minnesota, USA*)

27. Small-scale Model Test of a Frame–Shear Wall Structure 300
H. KRAWINKLER and B. WALLACE (*Stanford University, USA*)

28. Reinforced Concrete Beam-Column Subassemblages under Reversed Loading 311
E. C. CARVALHO and S. POMPEU SANTOS (*Laboratório Nacional de Engenharia Civil, Portugal*)

29. Use of Physical Silo Models 320
J. MUNCH-ANDERSEN and J. NIELSEN (*Technical University of Denmark*)

30. Static Tests on a 1:10 Scale Model of a Reinforced Concrete Chimney 329
A. CASTOLDI (*ISMES, Italy*), G. GIUSEPPETTI (*ENEL-CRIS, Italy*), L. RUGGERI (*ISMES, Italy*) and F. ULIANA (*ENEL-CTN, Italy*)

31.	Shear Reinforcement in Flat Plate Roofs A. ABDEL-RAHMAN, M. KHATER and M. NASSEF (*Cairo University, Egypt*)	338
32.	Static Test of Model RC Beams in Hydrostatic Pressure Surrounding N. K. SUBEDI and D. C. BELL (*University of Dundee, UK*)	353
33.	Comparative Analysis of Concrete Perforated Tubes Monolithically Cast and Made by Precasting M. MIHAILESCU, I. OLARIU, V. BUDIU and N. POIENAR (*Polytechnic Institute Cluj-Napoca, Romania*)	362
	Commentary R. N. WHITE (*Cornell University, USA*)	371
	List of Delegates	375
	Index of Contributors	377
	Subject Index	379

1 Microconcrete for structural model analysis

R. K. MÜLLER
Institute for Structural Model Analysis, University of Stuttgart, FRG

SUMMARY

Microconcrete is understood to be a material for preparing models for the structural analysis of concrete structures. With the aid of such models, the cracking and failure loads as well as the cracking distribution of reinforced concrete structures can be determined under realistic conditions. The microconcrete consists of portland cement and aggregates of river sand reduced to the model scale. Profiled steel wire, with diameters corresponding to the model scale, is used as reinforcement. Microconcrete models offer the advantage of significantly smaller size, lesser mass and lesser loading when carrying out tests as compared to tests on prototypes. Therefore, the models are less expensive to produce and simpler to handle.

1. INTRODUCTION

When designing reinforced concrete structures, it is now generally possible, using the analytical and numerical methods available, to calculate the purely elastic behaviour of a structure with sufficient accuracy. However, in the case of complex structures or when introducing new construction elements, the designing engineer will often be interested in the behaviour close to the failure limit, e.g., the capacity of a structure to change after partial failure and extensive deformations to a new stable condition. Due to the very intricate processes occurring during the plastic deformation of the multi-compound system reinforced concrete (cracking formation and redistribution of forces, complicated bond, non-linear elastic anisotropic material etc.), the condition close to the failure limit can only be numerically calculated with some difficulty and a fair amount of time and effort. Therefore, it has been found convenient to use model tests for these calculations. Realistic models, socalled real models, are indispensable for this purpose; experimental analysis using purely elastic models will normally prove insufficient /1/.

Real models are models which correspond during all stages of loading up to
the point of failure, to the load-bearing behaviour of the full-scale
structure; this is usually only possible when the real model is made of the
same material as the prototype. It is therefore necessary to use a concrete
whose aggregates are significantly smaller than the smallest model
dimension (usually the thickness). With model scales of 1 : 10 or less, up
to 1 : 15, only sand is still suitable as aggregate. This type of concrete
is referred to as microconcrete. With models made of microconcrete, it is
possible to ascertain, under conditions similar to the real ones, the
cracking and failure loads of reinforced concrete structures as well as the
cracking distribution. The advantages of microconcrete models are the
significantly smaller size, lesser mass and lesser loading required when
carrying out tests as compared to the analysis using prototypes. These
models are therefore cheaper to produce and simpler to handle. Microconcrete
models can be employed in research for the development of new modes of construction and for the design of complex structures in order to define the
safety against failure. Furthermore, the models are useful during the
training of students for demonstration purposes and practical studies.

The difficulties when analysing reinforced concrete structures, lie
essentially in the nature of the material. Reinforced concrete is known to
be a 2-component material consisting of a framework of aggregates, the
cement matrix and steel bars as reinforcement. In addition, there are
imperfections of various kinds. Therefore, reinforced concrete is
- non-isotropic,
  its physical qualities depend on direction,
- non-homogeneous,
  it has a macroscopic heterogeneous composition,
- non-linear elastic.

Since, on the building site, the method of production still heavily relies
on manual work, relatively great variations of the material characteristics
occur. They are caused by accidental influences such as air- and separation
zones along the reinforcement or by edge effects.

2. MODEL LAWS
The values obtained with the small-scale model are applicable to the

prototype when the physical conditions of similarity, i.e. the model laws, are fulfilled /2/. For real models, the $\sigma - \varepsilon$ -lines of the model concrete and of the model reinforcement must be affined to the prototype. Furthermore, a similar bond must exist. Otherwise a similar behaviour up to the point of failure cannot be expected.

Equal strains and lateral strain coefficients of model (M) and prototype (P) are basic conditions for strict similarity; additionally, all entities with the dimension of a stress must have equal scale factors:

$$E_{bV} = E_{sV} = \beta_{bWV} = \beta_{bZV} = \beta_{sZV} = \beta_{s0,2V} = \tau_{1V}.$$

Here,

$$E_{bV} = \frac{E_M}{E_P}, \text{ etc.}$$

signify the scale factors of the different entities.

Furthermore, the multi-compound model law must be observed for the composite material reinforced concrete. This requires that

$$E_{sV} = E_{bV}.$$

Since steel wires are used as reinforcement for the model as well as for the prototype, it follows that $E_{sV} = 1$. Thus, all scale factors for stresses and for Young's modulus must have the value 1. However, Young's modulus of concrete becomes smaller with decreasing maximum grain diameter, so that the requirement $E_{bV} = 1$ is difficult to realise for high-quality concretes of the prototype.

Therefore, one extends the similarity in relation to the cross section of the steel bars ($A_{sV} \neq 1_V^2$). The basic requirement of the strain equality $\varepsilon_V = 1$ must however be met; as a result

$$E_{bV} \cdot A_{bV} = E_{sV} \cdot A_{sV}.$$

The reduced model reinforcement area, with $E_{sV} = 1$, becomes

$$A_{sM} = A_{sV} \cdot A_{sP} = E_{bV} \cdot 1_V^2 \cdot A_{sP}.$$

This means, that by reducing the steel area of the model reinforcement, the rigidity is lessened and therefore the requirement of the multi-compound model law has indirectly been fulfilled.

3. CHARACTERISTICS OF THE MODEL MATERIAL
3.1 Model Concrete
When working with larger scale models, it is possible to remove the maximum grain of the standard concrete without significantly altering the behaviour of the material. Practical experience has shown, however, that this is only permissible up to a scale ratio of approx. 1 : 4. For this reason, analyses of reinforced concrete structures have up to now been mainly carried out with models nearly equal in dimension to the prototype.

For reasons of economy, for dealing with more extensive series and for the analysis of wide-span structures as a whole, a model material had to be found that makes it possible to produce real models on a smaller scale (up to approx. 1 : 15, depending on the size of the prototype).

The microconcrete developed in Stuttgart consists of portland cement and riversand aggregates reduced to the model scale /3/. It is known that the ratio of tension- to compression strength increases as the aggregate materials of the concrete mix are reduced in size. Thus, for the concrete mix used in most of the small-scale model tests, this ratio proved to be too large. As a consequence, the cracking loads were too heavy, and the crack widths too broad. It was therefore a primary aim when developing a microconcrete, to lessen the tensile strength without at the same time influencing Young's modulus or the compression strength. In order to develop suitable microconcretes for models that would correspond to the various quality grades of concretes according to German standard (DIN) 1045, the following parameters were changed:
- ratio of wet- to dry storage time at the beginning of the setting stage
- water-cement proportion
- aggregate-cement proportion
- nature and size of the aggregates
- use of additives.

For a greater reduction of the tensile strength of the model material, the larger size fractions of the aggregates are coated with a separating agent on a silicone base. This interferes in the contact region between cement stone and aggregate. The disturbed bond has an approx. equal effect on bond- and compression strength. This way, mixtures with low material strength can be produced.

Figure 1 Compression- and bending tension strengths as well as Young's modulus of the concrete mixes according to /3/

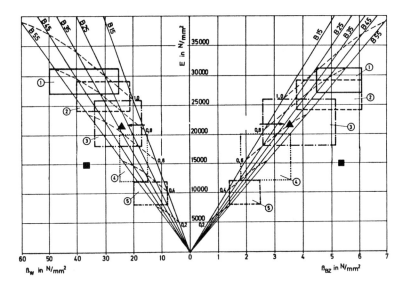

As shown by Fig. 1, the measures described produce rather good results. The graphical representation permits a comparison between the different model concretes and the concrete quality grades laid down as standards. On the ordinate, Young's modulus is shown, in the positive abscissa direction the bending tension strength and in the negative the cube strength are to be found. For the different qualities of concrete the German standard 1045 states different values for the cube strength, Young's modulus and the bending tension strength. A B 35, for example, has a cube strength of 40 N/mm^2, a Young's modulus of approx. 34000 N/mm^2 and a tensile strength of approx. 5,3 N/mm^2. The oblique straight lines represent the respective proper ratio between the tension- and compression strength of the standard concrete as compared to its corresponding Young's modulus. The dotted lines are the scale lines of the Young's modulus, resp. of the stresses. In order to examine if a model concrete corresponds to, e.g., a concrete of quality grade Bn 35, one transfers its Young's modulus, its tension- and compression strength onto the axes; after this, horizontal resp. vertical straight lines are drawn through the points on the axes. If these lines - both for the tension- as well as for the compression strength - intersect on the oblique straight line marked Bn 35, the model concrete corresponds to the quality grade Bn 35 of the prototype. From the position in relation to the dashed scale lines, one can recognize that this microconcrete has a Young's modulus, resp. a stress ratio, of 0,65. The rectangles contain all the values derived from the above mentioned measures. Precise information on the mixtures 1 to 5 is given in report no. 7 of the Institute for Structural Model Analysis, Stuttgart /3/. For example, the mixture 1 has a small water-cement factor whereas the mixture 5 has a very large one and, furthermore, contains additives.

In order to avoid variations of the concrete characteristics, the mixtures are produced to precisely stipulated formulations. The aggregates - reduced in size with geometrical similarity according to the grading curves of German standard 1045 - are dried and then added in individual grain fractions. This way, it is possible to achieve a precise composition of the grading curve even for small mixing amounts. A mechanical stirring device, on the principle of a paddle mixer, is used for producing the concrete mix. The test samples are compacted on the table vibrator with a vibration frequency of 50 cps and variable amplitude or with a small poker. In the case of intricate cross sections and complicated reinforcement, a good compaction result can only be obtained by means of an internal poker. The poker developed for this purpose has a diameter of 7 mm, a length of 55 mm and a weight of less than 10 g. It is driven via a thin flexible shaft at approx. 10.000 rpm.

3.2 Model Reinforcement

The smallest commercial diameter of structural steel amounts to 4 mm. For a scale length of 1 : 10, however, a model reinforcement with diameters between 1 and 3 mm is required. Steel wires with such diameters can only be obtained with smooth surfaces, and their ultimate strength lies between 650 and 950 N/mm^2 (bright-drawn wires) or 310 to 390 N/mm^2 (annealed wires).

Steel for construction, by comparison, has an ultimate strength between 500 and 600 N/mm^2. Since, according to the model laws, the reinforcement of the model must have a σ-ε-line identical to the prototype, the wires for the model have to be pre-treated:
- Either one increases the material strength of the annealed wires by cold forming;
- or one softens the bright-drawn wires by annealing them at a certain temperature.

Subsequently, ribs are squeezed into the pre-treated wires by cold rolling. Experiments have proved, that the geometrical form of these ribs does not have to resemble the prototype in order to achieve similar bond behaviour. The profiling device permits steplessly adjustable profiling degrees.

The profiling leads to a marked improvement of the bond behaviour; the bond stress becomes stronger with increasing relative translation of the wire in relation to the concrete (socalled shear bond). Additionally, the choice of the proper profiling depth makes it possible to obtain bond characteristics similar to the prototype.

3.2.1 Bond Behaviour

The best method for examining the bond behaviour is the pull-out test with short bond length. If one pulls at the embedded reinforcement bar, a bond stress τ_{1x} will operate which is a function of the length of the bar (Fig. 2). The greater the embedding length l_o, the more the local bond

Figure 2 Pull-out test with short and long bond length

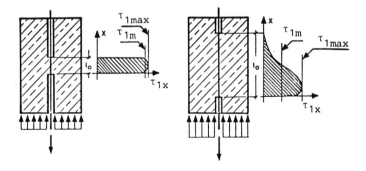

stress τ_{1x}, and equally the maximum bond stress τ_{1max}, will deviate from the mean stress τ_{1m}; only the latter can be calculated during the test as a quotient of force applied and embedded steel surface. For obtaining the maximum bond stress, l_0 must be kept rather short /4/, as shown on the left of Fig. 2. There, $\tau_{1m} \simeq \tau_{1max}$ applies. The shortest possible bond length is the distance between two ribs.

Fig. 3 represents the bond stress translation curves resulting from the pull-out tests. On the abscissa the measured translation of the wire end versus the concrete sample, and on the ordinate the bond stress divided by the compression strength are to be found. The dash-dotted line shows Koch's relation /5/ for the usual type of structural steel. The curve is converted to the model scale 1 : 5. One recognizes that these model wires at a length scale of 1 : 5 possess a similar bond behaviour. By changing the profiling degree, it is also possible to obtain the proper bond characteristics for other scales.

Figure 3 Diagram of bond stress vs translation

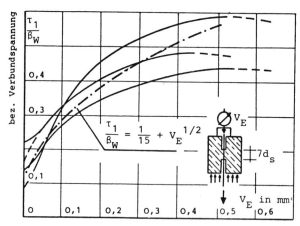

4. EXAMPLES

In order to demonstrate that the microconcrete technique developed in Stuttgart is suitable for conducting model tests with which the behaviour of the prototype can be predicted, various simple bodies were analysed.

The example of the published test results with reinforced concrete elements on a 1 : 1 scale was followed. For this,
- Falkner's tension bars on a 1 : 5 and 1 : 7,5 scale
- T-beams according to Ferry Borges under pure bending on a 1 : 11 scale
- beams under shear load on a 1 : 5 and 1 : 7,5 scale
- flat plates on a 1 : 6,45 scale subjected to punch-through tests (Fig. 4)

were produced as models and loaded analogous to the prototype. As far as amount of cracks, crack distances and crack widths were concerned, the behaviour of the microconcrete models essentially conformed to the prototype. Further details are given in the IMS-report no. 7.

Figure 4 Punch-through test on a flat plate with a 1 : 6,45 scale

REFERENCES

/1/ MÜLLER, R K, 'Entwicklung und Stand der Modellstatik', Der Stahlbau 52, no. 9, p. 264/268 (1982).

/2/ MÜLLER, R K, Handbuch der Modellstatik, Springer Verlag Berlin, Heidelberg, New York (1971).

/3/ SAUTNER, M, 'Ein Beitrag zur Entwicklung der Mikrobetontechnik', report no. 7, Institute for Structural Model Analysis, Stuttgart (1983).

/4/ REHM, G, 'Über die Grundlagen des Verbundes zwischen Stahl und Beton'. Deutscher Ausschuß für Stahlbeton, no. 138, published by Wilh. Ernst & Sohn, Berlin (1961).

/5/ KOCH, R, 'Verformungsverhalten von Stahlbetonstäben unter Biegung und Längszug im Zustand II', Thesis, Stuttgart (1976).

DISCUSSION

Peter Barr   Atomic Energy Establishment   UK

Over the past 6 years microconcrete targets have been constructed at the UKAEA Atomic Energy Establishment, Winfrith, Dorset, using a variety of aggregate gradings depending upon model scale. The maximum aggregate size used has been of nominal 10 mm diameter, but many mixes have had maximum aggregate sizes of about 2 mm diameter. In all about 300 models have been constructed, and work is continuing.

For each model about 70 test specimens are cast to provide compressive (150 x 300 mm cylinders, 150 and 75 mm cubes) and tensile strengths (Brazilian, four point bending and some direct tension), and pull-out discs (LOK-test) have been cast into models where potentially non-representative (eg water rich) zones could be avoided.

In general the models produced have no specific prototype but represent in general terms parts of nuclear reactor structures in which maximum aggregate sizes up to about 50 mm might be expected.

The measured values of tensile strength obtained by the Brazilian tests have been consistently about 10% of the compressive strength measured on cylindrical specimens. The results of four point bending beam tests have yielded tensile strengths about 25% higher than the Brazilian tests (ie about $12\frac{1}{2}$% of the cylinder compressive strengths).

We have been advised by our colleagues in the United Kingdom construction industry that these observed relationships between tensile and compressive strengths in our microconcrete are broadly the same as those expected in prototype concrete. Conversely, many users of microconcrete report tensile: compressive strength ratios of 1:5 unless special precautions are taken to reduce the tensile strength of the microconcrete. Professor Muller has described such techniques in his paper. I must presume that the cement: aggregate ratio could be a factor in producing these apparent differences. The cement:aggregate ratios used at Winfrith have been in the range 1:3 to 1:5 by weight.

Author's Reply.

At the Institute for Structural Model Analysis microconcretes with selected grading curves are being employed. Generally, aggregates of riversand are used with a maximum grain diameter of 4 mm resp. 2 mm; in special cases also 1 mm resp. 6,3 mm. During the time period 1980-1984 alone, approx 200 microconcretes with different mixing and storage conditions were produced. The aggregate:cement ratio has so far been between 3:1 and 4:1 (Winfrith 3:1 to 5:1). The corresponding tensile:compressive strength ratios lay between 1:7 to 1:10,5 and were measured on four point bending beams and cube specimens. For normal-weight concrete the tensile:compressive strength ratio for concretes B10-B55 between 1:5 and 1:8,3 (according to German Standard (DIN 1045), for practical building purposes also above 1:9. (When the results stated in the

'Comments' are converted to our testing methods, a ratio of 1:9,4 is obtained.)

It thus becomes apparent that our microconcretes show a result similar to that of the microconcretes mentioned in the 'Comments' which have a maximum grain of 2, resp. 10 mm diameter. In order to obtain these tensile-compressive strength results, no additives, apart from aggregate, cement and water, are required. Additives (eg Pluviol containing silicone) are used exclusively to reduce, where necessary, the tensile as well as the compressive strength to an equal degree, while keeping the same tensile:compressive strength ratio, so that the microconcretes can be better adapted to the results of the prototype.

2 Using models for studying reinforced fiber concrete behaviour

R. JOHN CRAIG, OSMANI LOY and CHARLOTTE RYDEN
New Jersey Institute of Technology, Newark, USA

ABSTRACT

Research has been previously done on the effects of steel fiber in concrete, both at New Jersey Institute of Technology and elsewhere. This paper explores the possibility of using small scale models which are made using microconcrete and gypsum models with fiber to mimic the prototype's behaviour. The mechanical properties of the models will be determined and compared to regular fiber concrete. The properties of compression strength, splitting tensile strength and flexural strength will be shown. Flexural and shear strength of the model reinforced concrete beams with and without fibers will be compared to the prototype reinforced concrete beams with and without fibers. Behaviour of actual scale modeling will be shown for: 1) a composite beam tests, 2) column tests, and 3) beam to column tests. There will be four different fibers used in this study. Problems of construction cover of reinforcing steel, size of fiber deflections, and cracking patterns will also be discussed. Conclusions will be drawn from the behaviour of models to actual prototype behaviour.

INTRODUCTION TO MODELS

Direct inelastic models are those that duplicate the behaviour of prototype structures through all loading stages up to and including failure(1). It is desireable to maintain a "table top" size in order to make the study easier. This usually means that a scale from 1:10 to 1:50 is desirable in order to avoid large scale loading facilities. Results should be obtainable in a short time and be able to be accurately reproduced.

Ultracal 30, manufactured by the U.S. Gypsum company, meets the above criteria. A two year study conducted at Cornell University resulted in the publication of "Small Scale Direct Models of Reinforced Concrete Structures" (1) which can be used as a handbook for working with gypsum models. Mix proportions of water, Ultracal, and sand, and necessary curing times are given to achieve various compressive strengths. Standard procedures for mixing, casting, and curing are described in reference from Cornell (1). Area of reinforcing steel is to be provided using the same reduction factor used for determining the model size. Bonding and yield strength are important considerations in choosing reinforcement material.

Also, in this paper microconcrete will be used for models. The Type III cement can be used to acquire strength at an early stage after casting. The modeling procedures are same as those used for the gypsum models.

The forms used for compressive strength and splitting tensile strength were plastic tubing, since they can be reused without change in their shape. 1 3/8" X 2 3/4" cylindrical plexiglass cylinders have been established as the equivalent to the standard 6" X 12" cylinderical mold used for determining f'_c for compressive strength and splitting tensile strength.

In order for the gypsum specimens to maintain a desired compressive strength, they must be sealed with shellac when that strength is reached. This prevents further loss of moisture. If not sealed, specimens compressive strength will continue to increase and they will become very brittle. The microconcrete and prototype concrete were cured identically as the beam specimens which were tested.

Basic tests conducted on microconcrete and gypsum concrete mixes have demonstrated that the behaviour of these mixes throughout the range of loading is very close to prototype concrete. Successful use has been made with gypsum concrete and microconcrete for more complex structures including t-beams, slabs, columns, and frames.

The advantage of model analysis as compared to a computer analysis is discussed in "Structural Modeling and Experimental Techniques"(2). Chief considerations in model studies are cost and time, which the analytical approach has on its side. Therefore, models are usually used when mathematical analysis is not adequate or feasible (2).

PROCEDURE

There were basically four different fibers which were used in this paper: 1) Short .125 in. carbon steel Fibercon Reinforcing fibers; 2) Dramix steel fibers ZP 30/50; 3) 3/4 Steel Ribtec fibers; and 4) ACE clipper No. 700 C. P. undulated staples.

The mix designs which were used for the prototype concrete; microconcrete; and gypsum concrete are as follows: 1) Prototype : 1098 lb/cy - coarse aggregate; 1600 lb/cy - fine aggregate; 404 lb/cy - water; and 674 lb/cy - cement; 2) microconcrete: 3900 lb/cy - fine aggregate; 913 lb/cy - water; 1522 lb/cy - cement; and 3) gypsum concrete : 1278 lb/cy - fine aggregate; 1278 lb/cy - ultracal; and 383 lb/cy - water.

The forms which were used for mechanical properties, flexural beams, shear beams, composite beam study, column study, and beam to column study were of different material. The plastic and steel molds were used for the gypsum concrete models; and the microconcrete and prototype concrete were cast in wood forms. The gypsum concrete was externally vibrated by shaking while the microconcrete and prototype were internally vibrated with an immersion type vibrator. The gypsum concrete was treated with a retarder to allow time for casting. The prototype concrete and microconcrete had a superplasticizer added to the mix in order to help the fiber concrete to be able to have better workability. Care had to be taken in casting in order to make sure that the fibers were cast correctly around reinforcing bars in the forms. The specimens which were made of gypsum were taken from their forms after two days and shellac was applied. The micro-

concrete and prototype specimens were taken from their forms at seven days
and tested around 28 days after casting. The details of the forms for
each test are as follows: 1) Gypsum concrete (with and without fibers)
were cast in 1" x 1" x 12" and 2" x 2" x 12" steel forms which had around
1:10 scale of the prototype. The molds for the compressive and splitting
tensile tests were made from plastic pipe and had a diameter to height
ratio of 1:2 with a diameter of 1-3/8"; and the microconcrete and prototype
concrete used various size wood forms depending on what type of behaviour
was desired. The molds for compressive and splitting tensile strength
were 6 in. x 12 in; 4 in x 8 in and 2 in x 4 in.

Testing Apparatus

Gypsum concrete beams were marked for loading points. Available
1 in. x 2 in. x 2 in. steel blocks with 1/2 in. diameter rollers were used for
supports. The series of beam tests were done having a bearing plate of
1 in. x 2 in. x 1/4 in. A Tininis Olsen testing machine, with a range of
0-50,000 Newtons, was used to apply the loads. Deflections were recorded
at 500 N and/or 1000 N intervals, using a deflection gage measuring to
1/1000 inches, placed at the beam center line.

The microconcrete models and the prototype specimens were tested in the
New Jersey Institute of Technology Structural Laboratory using loading
frames and rams of 8,000 lb., 32,000 lb., and 50,000 lb. capacities. The
loading frames were tied down to the test floor which has 40 tie downs
with a 50,000 lb. capacity each.

RESULTS

Mechanical Properties

The first series of tests are on the mechanical properties of the materials
being used such as compression strength, splitting tensile strength and
modulus of rupture. The typical compressive strength curves of the gypsum
concrete and prototype concrete can be seen in Figures 1 and 2 respectively. With the addition of fibers into the mix, there is an increase in
ductility and both model and prototype behaved similarly. There is only
a slight increase in strength because of the fibers. The splitting
tensile strength values for the gypsum models only increased slightly with
the addition of fibers as seen in Figure 3. The behaviour of the prototype

material is shown in Figure 4. The addition of fibers has somewhat more
influence on the prototype specimens. Figures 5 and 6 show the behaviour
of the modulus of rupture of the gypsum concrete and prototype respectively.
Again, the gypsum shows only a slight increase in the values of the modulus
of rupture with the addition of fibers whereas the prototype shows a large
increase when fibers are added.

Flexural Behaviour
Beam tests which show the flexural behaviour are shown in Figures 7 and 8
for gypsum concrete and prototype specimens with and without fibers. The
gypsum concrete specimens were 1 in. x 1 in. x 12 in. and 2 in. x 2 in.
x 12 in. The 2 in. x 2 in. x 12 in. specimens are shown in Figure 7. The
behaviour of Fibercon and staple fibers are also shown in the figure.
Figure 8 shows the behaviour of 7 in. x 120 in. prototype beams. The
prototype beams show more ductility than the gypsum concrete model beams
which showed less ductility when the fibers are added. Additionally, the
staple fibers show more ductility than the 1/4 in. Fibercon fibers. The
prototype specimens show a lot of benefit by adding fibers to the concrete.

Shear Behaviour
The three sets of experimental data shown in Figures 9, 10, and 11 show
the benefits of adding fiber. The gypsum models showed very little
increase in shear capacity as shown in Figure 9. The microconcrete specimens showed a 20 to 30 percent increase in capacity as shown in Figure 10.
This amount of increase has been what is normally accepted as reasonable
values by other researchers (3) and (4). When using the Dramix hooked
ended fibers, the capacity has been shown to be over 100 percent of
normal reinforced concrete beams. Figure 11 shows these results.

When looking at the test results shown here, influence of cracking and
fiber critical length (ℓ_c) are important. The fibers need to be representative in the models as they are in the prototype. The cracking behaviour
for the gypsum models are not representative. Figure 12a show the cracks
developed in the gypsum models and 12b show pictures of prototype models.
The cracks in the gypsum models are also hard to find because of the small
scale. Also, the critical length of the fibers is important. The use of
hooked ended and staples as fibers was to try to incorporate smaller
fibers as part of the modelling technique. The anchorage parts of these

smaller fibers was an attempt to develop the full load capacity of the fibers before pull out or failure of the structural members.

MODELLING AND BEHAVIOUR
Other model studies which have been investigated and have some behaviours which are worth mentioning are: One quarter scale of a composite beam; one quarter scale of a column; and one quarter to one half scale beam to column specimens.

The study of the one quarter scale of a composite beam showed that, because the cover of the reinforcing bars become smaller as a result of scaling the prototype, the confinement of the reinforcing bars was not as good as in the original prototype. The composite model specimen showed difficulties in casting because of the reinforcing bars present.

The one quarter scale columns were very difficult to cast because of its height and all the reinforcement present. The columns were cast in upright positions so that proper fiber alignment was obtained. The fibers present in a tied column also need cover on the outside of the column to prevent buckling. In order to get proper confinement, there has to be a proper amount of cover.

The beam to column specimens which were one half to one quarter scale showed good results. From this study, it was shown that having a larger section with more cover produced more benefits with having the fibers present.

CONCLUSIONS
The following are some of the conclusions obtained in this study:
1. The modeling of reinforced fiber concrete members should be kept to 1:4 and larger.
2. The model tests showed that the smaller fibers do not have enough critical length and straight fibers should be replaced by hooked ended fibers like the Dramix fibers or staples.
3. The cracks which are produced in the models are not comparable to the prototype behaviour.
4. The cover length of bars is important and may cause premature failures which might not occur in the prototype.

5. Model specimens could be used in a preliminary study to see the basic behaviours, but one quarter to one half to full scale models should be used to determine the overall behaviour of the specimens.

REFERENCES

(1) HARRIS, HARRY B, SABNIS, GASANAN, and WHITE, RICHARD, "Small Scale Direct Models of Reinforcing and Prestressed Concrete Structures," Cornell University, Dept. of Structural Engineering, School of Civil Engineering, (September 1966).

(2) SABNIS, GASANAN M, HARRIS, HARRY, WHITE, RICHARD, and MIRZA, M SAEED, "Structural Modeling and Experimental Techniques," Englewood Cliffs, N.J., Prentice Hall, Inc., (1983).

(3) BATSON, G, JENKINS, E, and SPATNEY, R, "Steel Fibers as Shear Reinforcement in Beams," ACI Journal, Pgs. 640-644 (October 1972).

(4) WILLIAMSON, G R, "Steel Fibers as Web Reinforcement in Reinforced Concrete," U.S. Army Science Conference Proceedings, Vol. 3, West Point, N.Y., (June 1978),

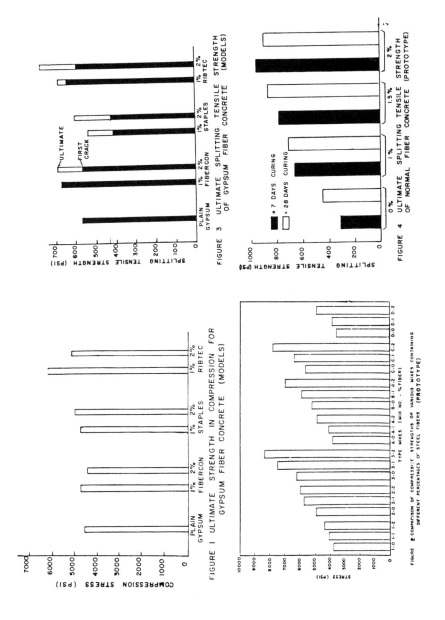

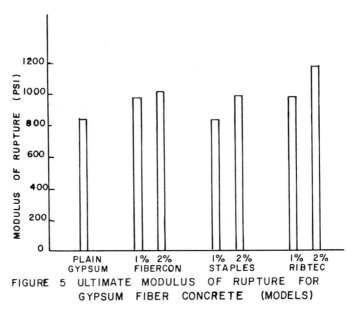

FIGURE 5 ULTIMATE MODULUS OF RUPTURE FOR GYPSUM FIBER CONCRETE (MODELS)

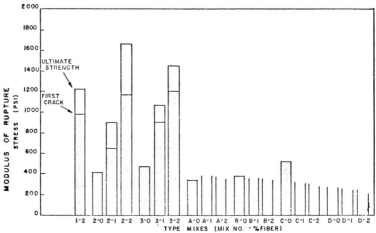

FIGURE 6 COMPARISON OF FLEXURAL STRENGTHS OF VARIOUS MIXES CONTAINING DIFFERENT PERCENTAGES OF STEEL FIBERS (PROTOTYPE)

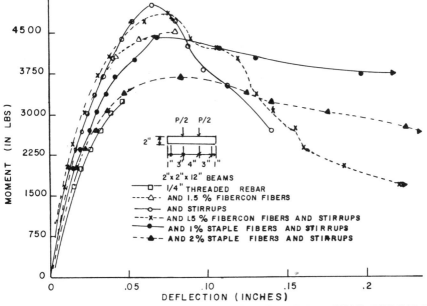

FIGURE 7 MOMENT DEFLECTION CURVES FOR SINGLE REINFORCED FIBER BEAM — REINFORCED FIBER GYPSUM CONCRETE

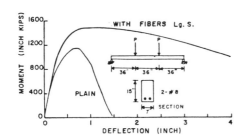

FIGURE 8 MOMENT DEFLECTION CURVES FOR SINGLE FIBER BEAM — REINFORCED FIBROUS CONCRETE. (PROTOTYPE)

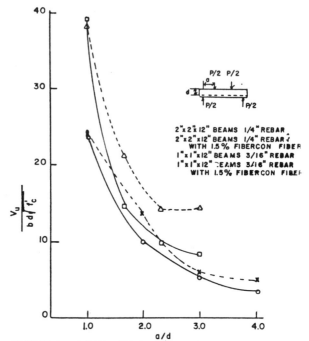

FIGURE 9   SHEAR BEHAVIOR OF REINFORCED FIBER GYPSUM CONCRETE BEAMS

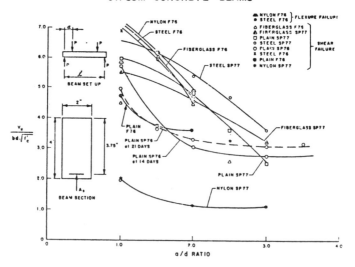

FIGURE 10   MODEL BEAM TESTS FOR FIBER CONCRETE SHEAR FAILURES

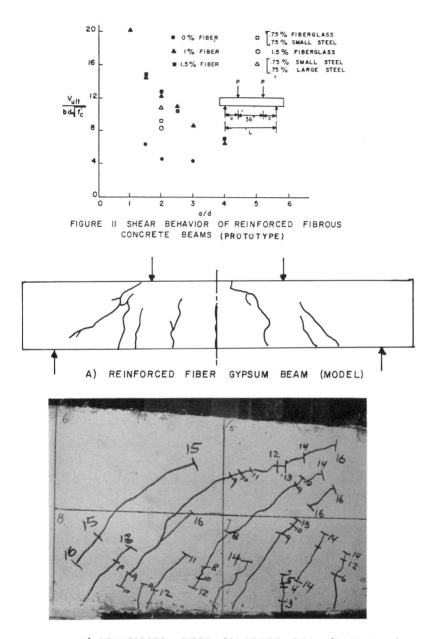

FIGURE 11 SHEAR BEHAVIOR OF REINFORCED FIBROUS CONCRETE BEAMS (PROTOTYPE)

A) REINFORCED FIBER GYPSUM BEAM (MODEL)

B) REINFORCED FIBER CONCRETE BEAM (PROTOTYPE)
FIGURE 12 COMPARISON OF CRACKING PATTERNS

DISCUSSION

Richard N White   Cornell University, Ithaca, N.Y.

We have experimented with fiber reinforced concrete models at Cornell, with application to the splitting strength of the end zones of prestressed concrete beams and to the strength of joints for frame structures. We used prototype size fibers made of brass coated steel. Problems were met in uniform mixing of the fibers, particularly with the lower water/cement ratio model concrete mixes. While the fiber reinforced model concrete frame joints showed increased strength and substantially larger ductility, relatively large amounts of fibers were required (4% by volume and higher). It would appear that more efficient joint designs could be made by utilising less steel but having it in the directions that it is most needed, that is, in directions of principal tension.

The fiber reinforced model beam end zones seemed to show better behaviour. Resistance to the concentrated splitting forces from anchored post-tensioning cables was greatly improved with the fiber reinforcement. Applications might be particularly attractive in situations where congestion of the other reinforcement made it difficult to place regular reinforcement to resist bursting forces. Or perhaps some combination of spiral reinforcement around the tendon, and fiber reinforced concrete for the entire end zone region, might be the optimal way of handling high local loads in concrete sections.

3 Development of a small aggregate concrete for structural similitude of slab–column connections

R. N. SWAMY and FAISAL M. FALIH
Department of Civil and Structural Engineering, University of Sheffield, UK

SUMMARY

The paper reports the development of a small aggregate concrete (sac) for model studies to simulate strength, cracking and deformation of slab-columns connections failing in flexure and punching shear. Tests are reported of 1/3, 1/4 and 1/6 scale models of such connections using mortar or sac as matrix, and mild or crimped steel, as reinforcement. The effect of scaling on cracking, mode of failure, and failure loads is reported. It is shown that whilst scaling effects are insignificant in flexural failures, they are critical in punching shear failures. One-third scale models gave the best structural similitude.

INTRODUCTION

Several model studies on slab-column connections subjected to different loading conditions have been reported. In some of these studies models have been used to define the shear and moment transfer at the connections in relation to Code provisions and analytical predictions. The effect of scaling has been investigated in other studies. In the investigation reported by Neth et al, the scale factors were computed based on the assumption that the failure loads of the 1/2 scale model were not influenced by size effects. It was found that reinforced concrete flat plate models at edge columns were influenced by size effects for models smaller than 1/4 scale while deflections were found to be unaffected by size effects. Model studies of bridge structures show that scale effects are not significant provided the specimens are carefully fabricated. This has since been confirmed by tests on 1/6 and 1/4 scale models to investigate punching failure at internal columns. However, it appears that the lack of scale effects is primarily due to the mode of failure being flexural.

The existing data, however, are not conclusive on the effect of specimen size on cracking, deformation, and on the mode of failure. Nor are the effects of scaling the two basic constituent materials - steel and concrete - clear on these aspects of slab - column behaviour. The results presented here form part of a wider study to develop a suitable model concrete and steel to investigate the total structural behaviour of slab - column connections, particularly in relation to punching shear.

Materials similitude is now recognised to be essential if structural similitude is to be achieved. The model concrete used here is based on the development of a small aggregate concrete (micro-concrete) compared to conventional micro-concrete (sand-cement mortar). Two problems are generally encountered in practice in using such micro-concretes - bleeding, which can have considerable effect on the cracking and deformation behaviour of the material, and the extent of scaling to be used for fine and coarse aggregates. These two difficulties have been largely overcome in the small aggregate concrete developed here - bleeding by replacing part of the cement by fly ash, and by modifying the mixing process, and the scaling, by using graded fine and coarse aggregates with specified maximum size. Model testing is further simplified by using the same matrix for different model scales. The choice of the small aggregate concrete (sac) itself is based on an extensive series of tests and a statistical evaluation of the effect of size of test specimen and scaling of aggregates on the strength and elasticity properties of the concrete. Five mixes (1 mortar + 4 sac mixes) were chosen to study the scale effects on compressive and split tensile strength. Based on these results, the mortar and one sac mix were chosen for further tests of modulus of rupture, elastic modulus, Poisson's ratio, effect of vibration time on compressive strength, shrinkage and swelling. The results of these tests will be reported elsewhere.

DEVELOPMENT OF SMALL AGGREGATE CONCRETE

Concrete Materials

The model concrete used in the tests was made from a blended cement of ordinary portland cement and fly ash. The fly ash was used to replace 50% by weight of cement directly. It had a specific surface of about 400 m^2/kg and a density of 2.17 gm/cm^3. Typically, the ash contained 56.2% silica, 26.2% alumina, 7.3% ferric oxide, 1.8% lime, 1.3% sodium oxide and 2.5% potassium oxide. A super-plasticizer was used with the mix as a water reducing agent, and to increase the flow characteristics of the concrete matrix.

The fine aggregate was washed and dried natural river sand, conforming approximately to grading zone 3. The coarse aggregate was graded crushed gravel.

Mix Design

The maximum aggregate size used in the prototype mix was 10mm. The maxmimum size of aggregate selected for the mortar matrix (used in the slab models of Series A and B) is 2.36mm (for all the three scales). In the small aggregate concrete (used for the slab models of Series C) a graded coarse aggregate with maximum size 6.7mm down to 2.36mm was added on to the mortar matrix. In developing the small aggregate concrete mix, the bulk density of the 10mm gravel, 6.7mm gravel and of the 2.36mm sand was measured: the proportions of the small aggregate concrete mix were designed such that the volume of aggregate used in it was similar to that used in the prototype mix since it is this rather than the gradation which strongly influences the properties of the model concrete.

The mortar mix and the small aggregate concrete used in the structural model tests were finalised after several trial tests in which the compressive and tensile (split test) strength were determined respectively from 50mm cubes and 50 x 100mm cylinders. The final mixes chosen are shown in Table 1a which also gives the details of the concrete mix used for the prototype slab tests.

To simulate tensile strength and ductility, the convenient parameter is the compressive-tensile strength ratio. This ratio was 12.97 for the prototype and 10.47 (Table 1b) for the mortar mix (used in Series A and B : 81% of that of the prototype mix); for the sac, this ratio was 12.39 (96% of that of the prototype mix).

Reinforcing Steel

The prototype slab had 10mm and 8mm diameter high yield steel deformed bars in tension and compression respectively. In this study only one bar diameter, namely 3.25mm, was used for all the three scales in all the three series. The reinforcement was used either as received as black-annealed plain mild steel wire (Series A) or as a crimped wire by cold drawing the former through a knurling machine (Series B and C).

The area of reinforcement was scaled down according to the following similitude equation:
$$A_m = \frac{S_\varepsilon}{S_\sigma} \cdot \frac{1}{S_L^2} \cdot A_p$$

Table 1a Mix proportions of prototype and model concrete

Mix	$\frac{F}{C+F}$	$\frac{S}{C+F}$	$\frac{G}{C+F}$	$\frac{W}{C+F}$	Admix. %	Slump mm
Prototype	0.3	1.8	2.2	0.47	0.6	103
Mortar	0.5	2.2	-	0.50	0.7	Collapse
SAC	0.5	2.2	1.8	0.50	0.7	Collapse

C - Cement; F - Fly ash; S - Sand; G - Gravel; W - Water

Table 1b Strength properties of prototype and model concrete

Mix	Comp. St. C, N/mm²	Tensile St. t, N/mm²	$\frac{c}{t}$
Prototype	43.2	3.33	12.97
Mortar	37.7	3.60	10.47
SAC	39.8	3.21	12.39

where A_p, A_m are the reinforcement area of the prototype and model respectively, S_ε is the strain scale factor, S_σ is the stress scale factor and S_L is the factor depending on the scale of the model.

The drawing and knurling process significantly altered the stress-strain behaviour of the original wire and made it resemble that of a cold drawn wire. The properties of the mild steel and crimped wire are shown in Table 2.

Table 2 Properties of reinforcing wire

	Plain wire	Crimped wire
Yield stress, N/mm²	198	348
Ultimate strength, N/mm²	325	350
Elastic modulus, kN/mm²	200	200
Yield strain, x 10^{-6} m/m	1230	2810
Ultimate strain, x 10^{-6} m/m	63000	21000

Table 3 Model concrete strength results

Slab No.	Comp.St., C N/mm²	Tensile St., t N/mm²	C/t
S1A, S1B	38.51	3.72	10.35
S2A, S2B	39.20	3.76	10.42
S3A, S3B	41.50	3.97	10.45
S4A, S4B	42.25	4.06	10.40
S5A, S5B	40.20	3.87	10.38
S6A, S6B	44.83	4.26	10.52
S7A, S7B	44.00	3.53	12.46
S8A, S8B	44.75	3.57	12.55
S9A, S9B	43.62	3.48	12.53

Mixing Model Concrete

One of the problems of using highly workable mortar or microconcrete mixes in model testing is that of bleeding during specimen manufacture. Bleeding can be reduced considerably and bond with steel improved by (1) using fly ash, and (2) by premixing the aggregate and the cementitious components with a proportion of the mixing water so that aggregate particles are coated with a layer of cement paste. The following mixing procedure was therefore adopted, and this was found to produce mixes of low bleeding.

For mortar:

| sand + W_1 | mixing 30 s | C+F | mixing 120 s | W_2 | mixing 60 s | Ad-mixture | mixing 60 s |

For small aggregate concrete:

| aggregate + W_1 | mixing 30 s | C+F | mixing 120 s | W_2 | mixing 60 s | Ad-mixture | mixing 60 s |

C - cement; F - fly ash; W_1 = 25% water by wt. (C+F); W_2 = Total water - W_1

DESIGN AND TESTING OF MODELS

The prototype was a flat plate structure with 4.0m centre to centre column spacing in both directions. The size chosen for the prototype tests was the part located within the negative moment region around the interior columns and inside the line of contraflexure. The size of the slab was thus 1800x1800x125mm with an average effective depth of 100mm. It was loaded through a 150x150mm square column. The slab was designed to be simply supported along the four edges with corners free to rise. This was considered as the best representation in a practical model of the actual boundary conditions in the prototype.

Three series of tests are reported: in each series, three model scales, namely, 1/3, 1/4 and 1/6, in which all linear dimensions were scaled, were tested. Each test was duplicated: eighteen model slabs were thus altogether tested. Series A used mortar mix with mild steel wire; Series B with the same mortar mix using crimped wire. Series C used the small aggregate concrete with crimped wire (Tables 1 and 2). The small scale models were tested in a specially designed frame, and loaded through the columns. A load cell was used to monitor the load.

Fabrication of Models

The two pairs of slabs for each scale and series were cast at the same time. All the specimens were compacted with a vibrating table. The moulds containing all reinforcement were initially half-filled with the mortar or sac matrix and compacted. Then the other half was filled and the vibration continued until all air bubbles were removed. The average time of vibration was about one minute. The moulds were then covered with polythene sheeting and left in the laboratory for 24 hrs. They were then demoulded and cured in a fog room at 20°C ± 2°C and 100% RH.

Four 50mm cubes and 4- 50x100mm cylinders were cast with each pair of slabs for compressive strength and split tensile strength. The slabs and the control specimens were tested at 28 days. The average values of the control strength tests are shown in Table 3.

Model Instrumentation

Extensive measurements of deformation were made during all the tests. The deflection was measured at the centre and, in the lateral and diagonal direction The vertical uplift at one corner was also monitored. The concrete strain was measured with a mechanical extensometer at several locations in the lateral and diagonal directions. Tensile steel strain measurements were generally made at four locations on all models with electrical resistance strain gauges. The rotation of each slab was measured, in the lateral and diagonal direction, with clinometer with an accuracy of 1 minute.

TEST RESULTS AND DISCUSSION

Extensive data were obtained from these tests on strength and deformation properties, crack patterns, mode of failure, ductility and energy absorption. Only the ultimate loads and the overall cracking behaviour are, however, discussed her

Cracking Behaviour

The overall cracking behaviour on the tension face of the prototype and of typical model slabs is shown in Figs. 1 to 4. The cracks were finer and more i number in all the models made with crimped steel i.e. Series B and C, compared t those made with mild steel reinforcement in Series A: the number of cracks also decreased as the model size was reduced (Table 4). For all the slabs, the cracks on the tension face were continuous, and no cracks generally appeared or the compression face of the slabs. Only in the 1/6 scale slabs of Series A anc B, the punching shear lines directly at the column sides appeared as a circle of 30 -40 mm diameter around the column.

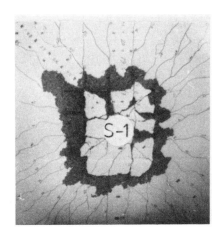

Figure 1   Prototype slab

Figure 2   Model slabs : Series A.

Figure 3   Model slabs : Series B.

Figure 4   Model slabs : Series C.

Figure 1

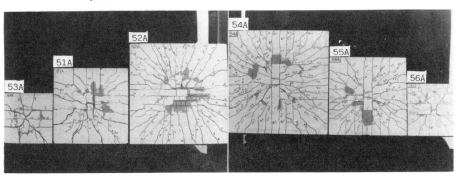

Figure 2   Figure 3

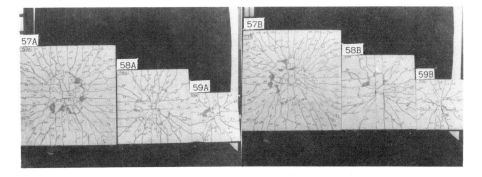

Figure 4

Table 4 Cracking and failure modes of prototype and model slabs

Series	Slab No.	Total no. of cracks on tension face	Average no. of side cracks	Expected mode of failure	Actual mode of failure
Prototype		–	10-12	Punching	Punching
Series A	S1A	31	7-8	Flex.	Flex-Punch
	S1B	31	7-8	"	"
	S2A	23	5-6	"	"
	S2B	18	4-5	"	"
	S3A	11	2-3	"	"
	S3B	11	2-3	"	"
Series B	S4A	42	10-11	Punching	Punching
	S4B	41	10-11	"	"
	S5A	28	6-8	"	"
	S5B	29	6-8	"	"
	S6A	16	4	"	"
	S6B	18	4-5	"	"
Series C	S7A	43	10-11	Punching	Punching
	S7B	39	9-11	"	"
	S8A	26	6-7	"	"
	S8B	25	6-7	"	"
	S9A	16	4	"	"
	S9B	16	4	"	"

The number and height of the cracks on the vertical sides of the model slabs also varied. These side cracks were more in number in Series B and C compared in those in Series A: also, they were slightly less long in Series C than in Series It was also noticed that none of the side cracks reached the upper edges in Series B and C, except for the 1/6 scale models where two cracks reached the upper edges of the slabs. Table 4 summarizes the total number of cracks on the tension face and on the vertical sides of all the model slabs.

Mode of Failure

The expected (i.e. from calculations), and actual modes of failure of the prototype and model slabs are also shown in Table 4. The prototype failed in punching shear as expected, and so did the slabs of Series B and C. The slabs of Series were expected to fail in flexure: these slabs did so after formation of the yield line patterns, and then, a punching failure occurred. The punching shear crack appeared for all the scale models at about 96-97% of the maximum load, and the failure was generally sudden at the maximum load. In the 1/6 scale models of Series A, however, this failure was slower and the load dropped to 92% of the maximum when the punching was completed.

Failure Loads and Design Loads

Table 5 shows the maximum test load carried by the prototype and model slabs, together with the expected maximum loads the models should have carried based on the prorotype maximum load and the model scale. The ultimate design loads in flexure and punching shear according to CP110 are also shown in the Table. The last two columns in Table 5 show the ratios of the maximum load to the expected maximum load of the model slabs.

The results show that there are small differences in the ultimate load capacity between the two slabs of each model scale in each series but these differences are generally small. It is clear that the results of model tests can be reproduced satisfactorily. The ratio of the ultimate design load to the maximum test load (col.6 Table 5) is nearly constant for all the model slabs of Series A, but higher than the expected value of 0.696. In Series B and C, the 1/3 scale models give the best correlation of these ratios to this expected value of 0.696.

Columns 7 and 8 of Table 5 show the ratios of the maximum test load to the expected failure load (based on the prototype load and the model scale). It is clear that there are no size effects in slabs of Series A mainly because the failure was in flexure which is in agreement with other findings. However, size effects are introduced in models of Series B and C because of the mode of failure in punching shear. This is in agreement with the findings of Neth et al.

CONCLUSIONS

The data presented here show that a small aggregate concrete containing a blended cement (50% OPC + 50% fly ash) and graded fine and coarse aggregates with specified maximum sizes can simulate adequately the properties of the prototype concrete. One-third scale models containing crimped wire and mortar or sac, showed the best cracking similitude. Scale effects for 1/3, 1/4 and 1/6 models are insignificant in flexural failures, but they are critical to punching shear failures. One-third scale models had negligible scale effects on punching shear failures; however 1/4 and 1/6 modes overestimated punching shear failure loads by up to about 6%-13% and 30-60% respectively.

ACKNOWLEDGEMENT

The authors wish to acknowledge the financial support received by the second author from the University of Basra, Iraq to carry out this project.

Table 5 Strength characteristics of prototype and model slabs

Series	Slab No.	CP110 Ultimate Design load, kN		Max. test load kN	Ult.design load / max test load	Max. test load / Expected test load	
		Flexure	Punch shear			value	average
Prototype		179.30	137.54	197.7	0.696	–	–
Exp. load : 1/3		19.922	15.282	21.967	0.696	–	–
Exp. load : 1/4		11.206	8.596	12.356	0.696	–	–
Exp. load : 1/6		4.981	3.821	5.492	0.696	–	–
Series A	S1A	13.575	17.555	20.883	0.841	0.951	0.938
	S1B			20.313	0.864	0.925	
	S2A	7.237	10.025	11.770	0.852	0.953	0.957
	S2B			11.878	0.844	0.961	
	S3A	3.342	4.464	5.391	0.828	0.982	0.967
	S3B			5.220	0.855	0.951	
Series B	S4A	18.285	15.849	21.836	0.726	0.994	1.019
	S4B			22.934	0.691	1.044	
	S5A	9.983	8.311	13.670	0.608	1.106	1.083
	S5B			13.101	0.639	1.060	
	S6A	4.611	3.733	7.259	0.514	1.322	1.327
	S6B			7.313	0.511	1.332	
Series C	S7A	18.332	15.849	23.271	0.681	1.059	1.045
	S7B			22.656	0.700	1.031	
	S8A	10.035	8.311	13.948	0.596	1.129	1.105
	S8B			13.350	0.623	1.080	
	S9A	4.612	3.733	8.325	0.448	1.516	1.552
	S9B			8.720	0.428	1.588	

DISCUSSION

M J G Connell   PSA   London   UK

In Table 5, Series C Slab Nos 9A and 9B results imply that for 1/6 Model Scale Punching Shear ultimate values are 30-60% greater than those for a Full Scale Test. Could you explain this result further please? It is not only due to Shear Angle difference 24° Full v 20° Model is it?

Author's Reply

There are three major factors responsible for the substantial over-estimation of the punching shear strength carried by the 1/6 scale models. These are

1. differences in steel area between the model and prototype
2. difference in tensile membrane action, and
3. scale effects.

The authors have discussed these three factors further elsewhere in the discussion, and would request Mr Connell to refer to these.

Author's Closure

The paper presented in this symposium is part of a wider study to develop simplified model materials and less complicated model techniques that are capable of simulating not only the ultimate strength but also cracking and deformation of the prototype member. There are three important modifications to conventional modelling that have been included in this study in developing material similitude. Firstly, for the mortar matrix of Series A and B, and the small aggregate concrete of Series C, the maximum size of aggregate (and the grading of sac) were kept constant for all model scales. In developing the small aggregate concrete, the volume of aggregate was controlled since it is this rather than the gradation which strongly influences the properties of the model concrete.

Secondly, the mixing technique was modified to control bleeding and this was achieved by pre-mixing the aggregates with a proportion of the mixing water so that the aggregate particles were coated with a layer of cement paste. This also helps the bonding within the model concrete and between the model concrete and the steel reinforcement. Several of the material properties such as compressive strength, compressive-tensile strength ratio, modulus of elasticity and Poisson's ratio of both mortar matrix and sac were investigated for different sizes of specimens. The stress-strain relationship of these materials is also being established.

Thirdly, the modelling technique was simplified by using a single wire diameter (ie 3.25 mm) for all the three scales. This of course resulted in varying steel percentages in the models compared to the calculated values from similitude considerations. These steel areas expressed as a percentage of the required value, were:

	1/3 scale	1/4 scale	1/6 scale
Tension	98-99%	96-100%	108-113%
Compression	92-99%	106-108%	79-123%

The improvements in the structural similitude of the slab models by using the small aggregate concrete of Series C are as follows:

1) The compressive-tensile strength ratio changed from 81% for Series A and B to 96% for Series C compared to that of the prototype.

2) The punching shear lines of the 1/6 scale model of Series C appeared directly at the column faces instead of a circle of 30-40 mm diameter around the column as happened for Series A and B.

3) The punching shear angle was about 20° for Series A and B, and about 25° for the slab models of Series C compared to about 24° of the prototype. Within the scope of this study, it appears that the punching shear angle is influenced more by the type of the model concrete and less by the scale of the model.

4) The deflection characteristics of the models were improved. For example, the ratio of the central deflection of the 1/3 scale model to the expected value was 1.86 for Series A, 1.44 for Series B and 1.07 for Series C.

5) The load carried by the tensile membrane action was as follows:

Scale	Ratio of the load carried by membrane action to maximum test load - %		
	Series A	Series B	Series C
1/3	33	32	31
1/4	33	34	32
1/6	49	46	41

Expressed as a ratio compared to that of the prototype, these values are:

Scale	Series A	Series B	Series C
1/3	1.32	1.28	1.24
1/4	1.32	1.36	1.28
1/6	1.96	1.84	1.64

For the 1/3 scale model of Series C this ratio was 1.25 - so again the benefits of the sac as a model material can be appreciated.

6) The modelling of the energy absorption was also greatly improved by the use of sac. For example, for the 1/3 scale model, the

actual energy absorption compared to the expected value was 3.12 for Series A, 1.52 for Series B and 1.17 for Series C.

7) The simulation of ductility could not be achieved partly due to the higher first crack loads in the models than those expected by scaling the first crack load of the prototype. It is obvious that the definition of ductility relating it to the first crack load needs modification.

8) Both cracking and the mode of failure were also better simulated by the sac.

The data presented in the paper show that scale effects are insignificant when the failures are flexural. In punching shear failures, the scale effects are again insignificant with the 1/3 scale models. With the 1/4 scale model, the ultimate load was over-estimated by 6-13% whereas with the 1/6 scale model, the ultimate strength was over-estimated by 30-60%. The large discrepancies observed with the 1/6 scale model are believed to be due to (1) differences in steel area (2) differences in tensile membrane action, and of course, (3) the scale effects.

4 The use of pushout tests to simulate shear stud connectors in composite beam construction

H. D. WRIGHT, P. W. HARDING and H. R. EVANS
University College, Cardiff, UK

SUMMARY

The proposed BS5950 design method for composite beams using through deck welded studs and profiled steel sheets uses a model test known as the 'push-out' test first used in CP117 to evaluate stud strength. This value may then be used for partial interaction design using limit-state methods. The test does not match the conditions occurring in the real beam and the code of practice underestimates actual strength and stiffness by approximately 25%.

From results of push-out tests and full scale composite beam tests it has been found possible to establish empirical correction factors to more closely match actual behaviour.

A more accurate method of design would entail using non-standard push out tests (including the profile steel sheet and through deck welded studs) to obtain design values. Such tests are described in this paper.

THE FLOORING SYSTEM

The use of profiled steel sheeting as permanent formwork has been common in America since about 1940 and soon after its use as tensile reinforcement in the slab was developed. Indentations in the steel sheet provide mechanical bond between concrete and steel. This so called longitudinal or slab composite action is often used with support steelwork utilising welded shear studs thus allowing transverse or beam composite action.

The strength of site welded studs has until recently been suspect and consequent many beams use shop welded studs. This, however, prevents the profiled steel sheets from being laid continuously.

Within the last 10 years, through deck stud welding has increased in reliability and is now the norm for office construction in America. The shear studs are welded using very high power forge welding machines (2400 Amp output). This power is sufficient to weld through the 0.7 to 1.2mm thick steel sheeting into the support steel below.

Thus the system can be used to its full advantage:- The steel sheeting can be laid continuously, over several supports, the studs are reliably site welded, and full two way composite action can be achieved.

The structural action of the beams using profile steel sheets differs in two respects from the more usual standard composite beam :
(1) The only contact between steel beam and concrete occurs at the stud (only a negligible amount of friction can occur between the steel beam and steel sheet).
(2) The presence of the profile reduces the amount of concrete in front of the stud taking the compressive shear bond forces.

THE DESIGN METHOD

With the introduction of this form of construction into Britain a new limit state design method has been proposed for the, soon to be released, BS5950. The method follows European recommendations and a design guide using these has been released by CONSTRADO[1].

The limit states applying to the design are based on ultimate strength and serviceability deflection at both construction (or wet-concrete stage) and the final condition.

However the ultimate moment and shear capacity in a final fully composite beam often exceed those required as construction stage beam size may well dictate strength.

To economise this condition a simplified linear version of "Partial Interaction Theory" proposed by Johnson[2] is included. The number of shear studs required for full composite action may be reduced by up to 50% providing that ultimate load and service deflections in this partial state are still satisfactory.

THE PUSH-OUT TEST

The stud strength used for these calculations is based on a model test known as the "Push-out" or "Push-off" test specified in CP 117[3]. Figure 1 shows the test specimen. The test provides not only ultimate strength but also stiffness characteristics although only strength is used in the design.

There is a numerical factor used to convert the test evaluated ultimate strength into a design value. This value 0.8 originates in CP 117 and is still used in the CONSTRADO guide. Yam and Chapman[4] derived this figure as a method of ensuring 125% interaction by the studs. The failure would therefore always be by flexural yielding of the beam rather than shear failure of the studs. For partial interaction when shear failure is expected the value 0.8 is inapplicable.

It can be seen from the standard test that the effects of including profile steel sheeting in the beam may change the characteristics of the test evaluated strength. This variation has been catered for by the reduction of stud strength dependent on profile steel sheet geometry. The reduction formula was derived by Fisher[5] and is based on tests carried out on American profiles.

TESTING PROGRAMME

The use of the CP 117 test and the reduction formula have been investigated by a series of CP 117 tests, full scale beam tests and non-standard push-out tests conducted by the authors.

24 standard CP 117 push-out tests have been carried out to determine stud strength of T.R.W. Nelson 19 mm studs in normal weight and lightweight concrete.

These values have then been used to design, to the CONSTRADO document, 8 full scale composite beams (see Figure 2). These beams utilise four British marketed profile steel sheets in various spans, rib orientations and concrete types.

To further investigate the effect of including profiled steel sheeting in composite beams 20 non-standard push-out tests which include the four British marketed profile steel sheets have been carried out. This latter test is as shown in Figure 3.

RESULTS

The standard tests confirmed that the T.R.W. Nelson stud was adequately strong. In all tests, failure was due to concrete crushing at the root of the stud. From the full range of 24 tests a design chart for stud strengths in varying concrete grades has been compiled[6].

The beam tests, designed on the basis of the standard push-out tests yielded ultimate strengths that were 25% in excess of the design method and stiffnesses that were 20% in excess. Again the failure mode in each case was by concrete crushing around the studs.

In fact so strong were the studs in the beam tests that a beam designed to provide 50% partial interaction failed at loads indicating 83%. Even when stud numbers were reduced below code requirements 73% interaction was the minimum obtained.

These high ultimate loads are unexpected as it was anticipated that the introduction of the sheet would have the opposite effect.

The results of the non-standard push-out tests confirmed this assumption. The introduction of the deck in the test generally reduced strengths and stiffnesses. It can be seen from Fig. 3 that while the ultimate load may be reduced by only 10% in somecases the introduction of certain decks and the position of the stud within the trough of the deck may cause much greater reductions and variations in stiffness.

It was also noted that lower stud strengths were apparent even when the reduction formula showed that no strength reduction was required. In fact only one deck 'Metecno' requires stud strength reduction according to Fishers formula.

SUITABILITY OF THE STANDARD PUSH-OUT TEST

There are two areas where the use of CP117 standard push out tests appear questionable in limit state design of composite beams using profiled steel sheeting and through deck welded studs. The first area is in the choice of numerical factor used to establish the design strength of the stud from ultimate test data. The second is in the variation of stud strength due to the inclusion of the deck not catered for by the reduction formula.

(i) As the design method follows limit-state philosophy it would appear sensible to include a partial safety factor for materials in the conversion of push-out test ultimate strength to a design value. It has been shown that concrete crushing is the failure mode of studs and the CP110[7] factor of 0.67 would therefore allow for variations in concrete quality in the slab as compared to the laboratory controlled test mix. (No material factor was used in beam test comparisons as the beam was also cast in the laboratory and concrete quality was ensured).

It has been clear from the beam tests that the push-out test results, when used with the design method underestimate strength and stiffness. The extent of this underestimation has led the authors to believe that the stud in the beam, by virtue of the compressive restraint afforded by the concrete slab, is stronger than the stud in the push out test.

This has led to the proposal of a "correction factor" similar to the 0.67 used in CP110[7] to evaluate allowable bending compression from compression only cube results.

The linear partial interaction design method relates beam strength directly to partial interaction by the formula :

$$M_{ULT} = M_{PLAST} + \frac{N}{N_f} (M_{COMP} - M_{PLAST})$$

where

M_{ULT} = Ultimate moment of resistance of the composite beam
M_{PLAST} = Ultimate moment of the steel beam alone
N = Number of studs required
N_f = Number of studs required for full composite action
M_{COMP} = Ultimate moment of resistance of the beam assuming full interaction

From the range of tests we have values for M_{ULT} and N. From the design method we can use M_{PLAST} and M_{COMP}. It is a simple matter to determine N_f for each beam.

But according to the design code

$$N_f = \frac{F_s}{\gamma_{TS} P_d}$$

where F_s = the force in the steel beam

P_d = The ultimate shear strength of a stud
γ_{TS} = A correction factor

Thus a correction factor applied to stud strength may be predicted by comparisons of test ultimate load and unfactored design method ultimate load.

From the tests conducted to date a lower bound figure of 1.8 is indicated. In using this figure it must be remembered that in addition to correcting for the increased stud strength in the beam compared to the test, it may also be compensating for effects such as strain hardening, concrete slab strength and variations from true non-linear partial interaction.

Despite improvement in ultimate strength prediction it is also important to check service deflection. Service deflection is calculated from the following formula (proposed by Johnson).

$$\delta = \delta_f + \frac{1}{2}(\delta_s - \delta_f)|1 - (N/N_f)|$$

where
δ = deflection of the beam
δ_f = deflection assuming full composite action
δ_s = deflection of the steel beam alone
N/N_f = as before

It can be seen that this prediction is based purely on stud strengths. No account of stud deformation is taken. The factor 1.8 should therefore be equally applicable. However the result of using this factor can be seen on Fig. 4. The design method may overestimate stiffness when used with the actual degree of partial interaction (simulated by the use of the correction factor). The formula is only safe when used in codes of practice by underestimating partial interaction.

It is therefore prudent to use a lower correction factor for deflection calculation. This appears to be dependent on deck geometry and from the range of tests carried out a lower bound figure of 1.4 appears reasonable. This is shown on Fig. 4.

(ii) Given the underestimation of strength and stiffness by the design method it would appear that the CP117 standard push out test is suitable for stud strength prediction.

The variation in stud strength shown during the non-standard push out tests proves however that this would be unwise. None of the beams tested utilised studs to the front of the trough and this is fortuitous as a 40% reduction in stud strength may well have caused much lower beam strengths and stiffnesses.

There are two approaches to the solution of this problem. The first is to evaluate a new strength reduction formula more applicable to British deck geometry. The second is to adopt the non-standard push-out test as a method of establishing stud strength when profile steel sheets are used. In America where the number of manufacturers (each producing several decks) is in excess of 20 this latter suggestion is prohibitive. However in Britain where the number of manufacturers are substantially smaller the case is stronger.

CONCLUSIONS

The design method proposed for the new limit-state code of practice BS5950 underestimates both strength and stiffness of composite beams profiled steel sheets and through deck welded studs.

The design strength of the studs used in both beam strength and stiffness design is determined by the CP117 standard push-out test.

The authors believe that the factor of 0.8 used to obtain design stud strengths from the CP117 test may be modified to 1.8/1.4 to take account of increased strength/stiffness of the stud as found from full scale beam tests. This factor should be linked with a partial safety factor for materials of 0.67 to match current limit state design philosophy.

It is also suggested that the CP117 test should be modified to include the appropriate profiled steel sheeting thus more accurately modelling the actual beam.

ACKNOWLEDGEMENTS

The authors would like to thank T.R.W. Nelson for funding the above research work.

REFERENCES

(1) CONSTRADO, Steel framed multi-storey buildings. Design recommendations for composite floors and beams using steel decks. Section 1 Structural (October 1983).

(2) JOHNSON, R P, Composite Structures of Steel Concrete, Vol. 1, Granada (1975).

(3) CP117, Composite Construction in Structural Steel and Concrete, Part 1 : Simply-supported Beams in Buildings, British Standards Institution, 1965.

(4) YAM, L C P, and CHAPMAN, J C, "The inelastic behaviour of simply supported composite beams of steel and concrete", Proc. I.C.E. 41. 651-684,(Dec.1968).

(5) FISHER, J W, "Design of composite beams with formed metal deck", Eng. J. Amer. Inst. Steel Constr., 7, 88-96, (July 1970).

(6) T R W NELSON, Weld-Through Deck Application and U.K. Design Data Manual.

(7) CP110. "The structural use of concrete", British Standards Institution, (1972).

Figure 1   CP117 Standard push-out test specimen

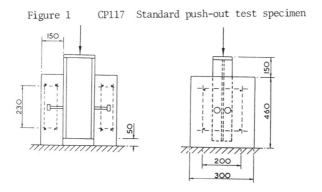

Figure 2   Non-standard push-out test specimen

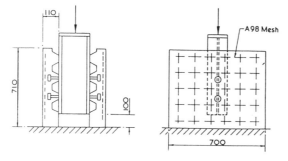

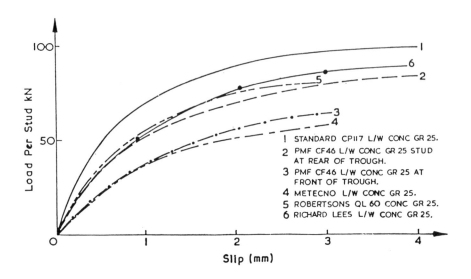

Figure 3 Comparison of standard push-out test with non standard tests using various decks and stud position

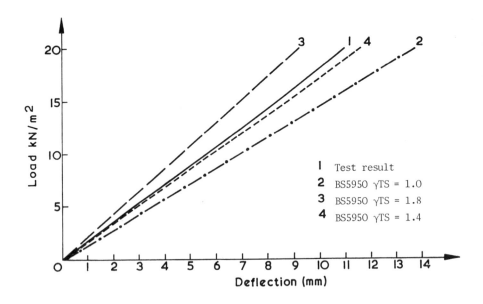

Figure 4 Stiffness of Richard Lees Holorib deck on 203 x 213 x 25 U.B. with 10 No. 19mm studs (span 4.5m breadth 1.5m)

TEST PROGRAMME

Concrete	Deck	% composite action	direction of corrugation	span (mm)	test date
Lightweight	CF 46	80 - 90%	T	4.5	Aug. 83
Lightweight	Holorib	80 - 90%	T	4.5	Aug. 83
Normal weight	Holorib	80 - 90%	T	4.5	Oct. 83
Lightweight	Metecno	50% approx.	T	5.5	May 84
Lightweight	QL60	50% approx.	T	6.0	May 84
Normal weight	CF46	50% approx.	T	6.0	Feb. 84
Normal weight	Metecno	100%	P	6.0	Jan. 84
Lightweight	Metecno	100%	P	6.0	Jan. 84

This table gives details of the 8 full scale beam tests carried out by the authors. The results of these tests were used to obtain the test correction factors detailed in the paper.

5 Small scale modelling techniques for investigating the failure of a bin structure

RICHARD N. WHITE
Cornell University, Ithaca, New York, USA

SUMMARY

A 1/12 scale plastic model study of the behavior of a steel bin structure is described, with emphasis on similitude requirements, model fabrication methods, model loading, and determination of gage stiffening effects.

INTRODUCTION

Potential elastic instability modes of a steel bin structure were studied by tests on a small scale plastic model bin. The physical modeling approach was chosen because of the unusual difficulties met in attempting to determine potential instability modes in thin-shell structures.

Although the prototype structure was made of steel, the modeling techniques use in the study are applicable to concrete shells. Hence this paper will focus on these techniques, which include special fabrication methods for the plastic sheets, an unusual combination of materials for loading the model bin, and determination of the stiffening effect of the strain gages mounted on the very thin plastic sheeting. Ref. (1) gives more details.

The prototype (full-scale) bin geometry is shown in Fig. 1. The enclosed volume of the structure was formed by a cylinder and a truncated cone. Twelve equally spaced wide flange steel columns, terminating at the underside of the cone, were welded intermittently to the interior surface of the cylinder. Plate thickness are shown in Fig. 1. The prototype volume was about 19,000 ft^3 and design capacity was 550 tons, or 60 lb/ft^3.

SIMILITUDE CONSIDERATIONS

The pertinent set of dimensionless ratios to satisfy similitude for the elastic behavior of the bin are:

$$t/\ell,\ \sigma/E,\ \nu,\ \varepsilon,\ \rho\ell/E,\ \phi,\ \phi',\ \text{and}\ \beta \tag{1}$$

where t = plate thickness
- ℓ = characteristic length such as bin height or diameter
- σ = stress in plates and columns, or pressure of bin material on the bin walls
- E = Young's modulus of plate and column material
- ν = Poisson's ratio of plate and column material
- ε = strain in plates and columns
- ϕ = angle of internal friction of contained material
- ϕ' = angle of sliding friction of contained material on the bin wall
- β = compressibility factor defined from $(\rho/\rho_0) = (\sigma/\sigma_0)^\beta$ where ρ_0 and σ_0 are base values of density and applied stress, σ is the actual applied stress, and ρ is the resulting density, all in terms of the contained material.
- ρ = density of contained material

Equating these dimensionless ratios for the model and prototype structures gives design and operating conditions for the model. The first four dimensionless ratios lead to the usual relations (see Ref. 2) for geometric scaling, stress scaling, and equality of strains and of Poisson's ratio for model and prototype. The fifth ratio gives the scale factor for density of contained material,

$$\rho_m = (s_\ell/s_E)\ \rho_p \tag{2}$$

where the scale factor s_i for quantity i is defined as i_p/i_m, with the subscripts p and m denoting prototype and model, respectively. The last three dimensionless ratios express the fact that the contained material used in loading the model should have the same friction angles and compressibility factor as the contained material in the prototype.

By substituting stress σ for modulus E in the dimensionless ratio used in arriving at Equation 2, we obtain the relation

$$(\sigma/\rho\ell)_p = (\sigma/\rho\ell)_m \tag{3}$$

According to Johanson (3), the major principal stress σ_1 in a bin is

$\sigma_1 = G_1 \rho D$ where G_1 is a dimensionless geometry factor dependent upon the shape of the bin, ρ is the contained material density, and D is a characteristic length such as the bin diameter. This gives the same dimensionless ratio as obtained from Eqn. 3.

MODEL MATERIALS

Plastic Sheeting: Rigid planished vinyl sheeting and polyvinylchloride (PVC) columns were used to model the plates and columns of the prototype bin. Vinyls are often preferred over acrylic plastics such as Plexiglas for model construction because of more uniform thickness and lower creep tendencies when loaded; in addition, PVC wide flange column sections are available from model parts supply companies. The sheeting had a modulus of about 425,000 psi, and Poisson's ratio was about 0.35.

Contained Material: Substituting values of $s_\ell = 12$ (this geometric scale factor will be justified in the next section) and $s_E = (29,500,000/425,000) = 69.4$ into Equation (2) gives a model loading material density of $\rho_m = (12/69.4) \rho_p = 0.173 \rho_p$. Hence the design density for the contained model material is $\rho_m = 0.173(60) = 10.37$ lb/ft^3. Three different densities were used in the five model tests: Test #1 - 6.8 lb/ft^3, tests #2,3 - 12.7 lb/ft^3, tests #4,5 - 15.6 lb/ft^3. Thus the actual model loads were either 22% or 50% over design loads for the last four tests.

The basic loading material for the model was ground vermiculite with a density of about 6.8 lb/ft^3. The higher density values were achieved by mixing in the proper proportions of precooked dry rice, and rice plus cracked corn, respectively, to achieve 12.7 and 15.6 lb/ft^3. All mixes had a natural angle of repose of 30 to 35° as compared with 32° for the powdered rock in the prototype. Arriving at this rather unusual combination of loading materials was perhaps the most difficult part of the modeling study.

The angle of sliding friction against the bin wall was not measured for the model material, but it was very small because of the "slippery" surface of the plastic and hence produced scaled bin pressures at least as high as the prototype material with its designated ϕ' value of 22°. The final material factor, the compressibility coefficient β, was ignored in this modeling study inasmuch as neither the model or prototype contained material had any appreciable compressibility effects.

DESIGN AND CONSTRUCTION OF THE MODEL

A geometrical scale factor of $1/s_\ell = 1/12$ was chosen to provide the best match of scaled prototype plate thicknesses to available sheet plastic. Table 1 gives a summary of the prototype plate thicknesses, the "theoretical" model thicknesses obtained by dividing the prototype quantities by 12, and the actual model thicknesses. In all cases the actual thicknesses are less than required by similitude theory.

Table 1 - Plate thicknesses, inches

Plate	#1	#2	#3	#4	#5
prototype	3/16	5/16	3/8	1/4	0.135
prototype/12	0.0156	0.0260	0.03125	0.0208	0.01125
model	0.010	0.025	0.030	0.020	0.010

The cylindrical portion of the bin was made up of a single 48.17 in. by 91.11 in. sheet of 0.010 in. thick rigid planished vinyl reinforced with two strips of thicker plastic (0.015 and 0.020 in. thick) as shown in a developed view of the bin wall in Fig. 2. The 6 in. and 4 in. wide strips of vinyl were laminated to the base sheet with Multi-Purpose Spray Adhesive (3M Company), and the ends of the strips were located to provide a staggered joint when the closed cylinder was formed. Laminating was done with the plastic sheets curved around a cylindrical metal forming mold to prevent the development of residual stresses and imperfections in the bin wall. The vertical joint in the cylinder was made with an aluminized mylar tape placed both inside and outside the cylinder.

PVC columns were machined to the proper scaled dimensions and fitted with an inclined end cap. The columns were cemented into the cylinder with 1/4 in. long connections of Vinyl Adhesive SC-203 (Cadillac Plastics Co.) spaced 1 in. on centers on each edge of the flange-to-cylinder contact surface. These connections modeled the 3 in. fillet welds spaced 12 in. on center in the prototype bin.

The truncated cone was formed from 0.020 in. thick vinyl and reinforced at its lower end with two Plexiglas rings. This end detail is a conservative representation of the large, stiff unloading device bolted to the bottom of the prototype cone. The model cone edges were joined with reinforced mylar tape. The cone was then positioned to rest on the inclined end caps of the twelve columns and fastened to the interior of the cylinder with Vinyl Adhesive SC-203. The model was completed by fabricating a 0.020 in. thick curved angle for the base of the structure, and cementing it to the columns with Adhesive SC-203. No cover was placed on the model and the stiffening angle at the top of the bin was omitted. The completed model bin is shown in Fig. 3, fastened to a wooden base ring made of two thicknesses of 3/4 in. plywood supported on twelve 2 by 4 in. wooden legs one under each column of the bin.

INSTRUMENTATION

Strains were measured with 32 electric resistance strain gages located near the cone-cylinder junction; seven 3-gage delta rosettes and eleven single element gages were used. All gages were Micro-Measurements 120 ohm foil element, epoxy backed gages. The gages were wired into Wheatstone quarter-bridge circuits with a similar unstressed dummy gage used for temperature compensation. Dial gages were used to measure displacements.

TEST RESULTS

Introduction: Two sets of strains resulting from full load were obtained. The first strains were those measured during the incremental loading (15 to 20 lb increments) over a time period of about 2 hours and thus included creep effects. More accurate strains were obtained by determining the strain changes during the 2-1/2 minute unloading of the model bin. Fig. 4 shows the loaded bin.

Gage Stiffening Effect: The metal strain-sensing material and the epoxy backing of the strain gage have elastic modulus values higher than the base value for the plastic. This effect was investigated by conducting modulus measurements on plastic specimens both with and without single element strain gages cemented on the specimens. Specimens instrumented with strain gages and loaded with axial tension stresses in the range of 400 to 800 psi gave apparent modulus values as follows: $t = 0.01$ in., $\bar{E} = 728,000$ psi; $t = 0.015$ in., $\bar{E} = 544,000$ psi; and $t = 0.02$ in., $\bar{E} = 493,000$ psi. Additional tests were done on specimens 0.010" and 0.020" thick, but with no strain gages attached, on an Instron testing machine. The modulus values for both thicknesses were within 1% of 425,000 psi. True

strains in the model bin were estimated multiplying the observed strains by the
$\bar{E}/E = E/425,000$, or for the three thicknesses of plastic used in the model (0.01,
0.015, 0.02 in.), the adjustment factors were 1.71, 1.28, and 1.16.

Strains and Stresses: Maximum hoop strains in the cylinder measured with gages
22, 23, 25, and 26, and adjusted for local stiffening effects, are plotted in
Fig. 5a. They are joined by a curved line drawn to represent the probable variation of compressive hoop strains in this critical region of the bin. A peak compressive strain of 0.000520 (520 microstrain) is estimated at the intersection.
The strain distribution indicates that a section of the cylinder slightly greater
than 2.5 inches in depth, as shown in Fig. 5a, acts as part of the compressive
ring at the cone-cylinder intersection.

Hoop strains in the cone as determined from gages 27 and 29 are plotted in Fig.
5b. The strain of -503 at the top of the cone compares well with the strain of
-520 in the cylinder at the junction. These results also indicate that compressive strains exist in the top 1.4 inches of the cone, with a transition to tensile hoop strain occurring at this distance from the intersection. The scaled
peak stress in the prototype at the cone-cylinder junction would be -24,300 psi,
which is below the elastic limit of the steel plate material. Meridional tensile
stress in the cone was compared with simple equilibrium stresses calculated
assuming zero friction between the bin material and the cylinder. The measured
stress was about 20% higher than that provided by equilibrium calculations, but
some differences were expected because bending stresses of undetermined value act
in the cone.

Deformations and Buckling: The cylindrical portion of the bin (from the tops of
the columns to the base) acted compositely with the twelve columns in carrying
vertical compressive stress produced by the weight of the contained material.
Because this cylinder was extremely thin (0.010 in. in the model) it tended to
buckle from meridional compressive action as the bin was loaded, with visible
buckling at 50 lbs load and extensive buckling at the design load of 115 lbs
as shown in Fig. 6. The photo shows reflected light patterns which greatly
amplify the true buckling deformations. There was no detectable overall
instability in the compressive ring region at the cone-cylinder intersection.
This behavior mode was the key quantity to be assessed from the model study since
ring instability would have been critical in controlling the strength of the bin.

CONCLUSIONS

The following conclusions are made for the model bin loaded with 1.5 times the design loading:

a. laminated rigid planished vinyl sheeting is an excellent material for fabrication of variable thickness small scale models of shells.
b. a highly unusual combination of loading materials was needed to meet density and internal friction angle similitude requirements.
c. the model bin carried 1.5 times design load with no instability of the compressive ring area at the cone-cylinder intersection.
d. maximum strains occurred in the cone-cylinder intersection region, resulting in a maximum predicted prototype bin stress of slightly over 24,000 psi (compressive hoop action).
e. there was extensive buckling of the cylindrical skirt area below the cone-cylinder intersection, but this action did not lead to any adverse overall structural behavior of the bin. The skirt continued to function effectively in providing adequate lateral restraint against weak axis buckling of the twelve supporting columns.

REFERENCES

(1) WHITE, R N, 'Model Study of the Failure of a Steel Bin Structure,' Proceedings of ASCE/SESA Session on Physical Modeling of Shell and Space Structures, New Orleans, October 1982.

(2) SABNIS, G M, WHITE, R N, MIRZA, S M, and HARRIS, H G, Structural Modeling and Experimental Techniques, Prentice-Hall, 1982.

(3) JOHANSON, J R, 'Modeling Flow of Bulk Solids,' Powder Technology, 5(1971/72), pp. 93-99.

Fig. 1 Prototype bin dimensions

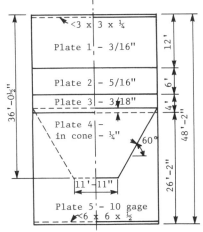

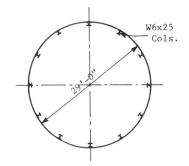

Fig. 2 Developed view of model bin cylinder

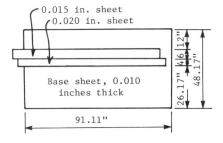

Fig. 3 Model Bin

Fig. 4 Loaded Bin

Fig. 6 Buckling of Cylinder Below Cone-Cylinder Junction

Fig. 5 Strains at cone-cylinder junction

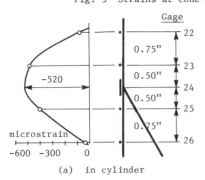

(a) in cylinder

(b) in cone

DISCUSSION

Jorgen Nielsen   Technical University of Denmark.

The following remarks concern partly the similitude consideration, and partly the accuracy.

SIMILITUDE CONSIDERATIONS

The model law developed in the paper is valid, if the list of parameters in question is the one showed in the paper, and if the major principle stress can be obtained by the relation

$$\sigma_1 = G_1 \rho D$$

However, this equation is based on several assumptions, which are not listed in the paper. Later in the paper it appears that the compressibility of the particulate material is ignored.

Another way of obtaining the result could be to assume plastic behaviour of the particulate medium, covered by the Coulomb's yield criterion (without cohesion).

This leads to the following equation for the medium:

$$s_\phi = 1, \quad s_{\phi'} = 1$$
$$s_\sigma = s_\rho s_\ell$$

Hereby the loads on the structure are formed without restrictions on the strains, as we have assumed plastic behaviour.

To study elastic buckling of the structure we need s = 1. With geometrical similitude this calls for

$$s_E = s_\sigma$$

which combined to the above gives

$$s_\ell \cdot s_\rho = s_E$$

This is the equation used in the paper as part of the discussion of model materials.

The same could be obtained, if the same basic argumentation is used to change the list of dimensionless ratios mentioned in the paper: $\rho\ell/E$ should be substituted by $\rho\ell/\sigma$ and   should be omitted.

ACCURACY OF LOAD DISTRIBUTION

Substitution of one material with another according to the model law is possible only if all important parameters are covered by the model law. In this case only ϕ, ϕ and   are taken into account. Results from soil mechanics and silo model test have revealed that anisotropic behaviour and stress history play very important roles, as they make the shape of the particles and the filling procedures important parameters. In particular it is found that small changes in these conditions may course a change in the flow pattern with a big change of pressure distribution as a consequence (1).

This means that the stress distribution obtained may be good enough for the structural analysis in this case, but in general and especially for load research purposes the accuracy seems doubtful.

(1) J Munch-Andersen and J Nielsen: Use of Physical Silo Models. Seminar on "Design of Concrete Structures - The Use of Model Analysis", 29th and 30th November, 1984. Watford. England.

6 Transient loading of structural concrete

H. K. CHEONG, A. H. AL-SHAIKH, N. N. AMBRASEYS and S. H. PERRY
Imperial College of Science and Technology, London, UK

SUMMARY

Cyclic static tests have been carried out on small specially reinforced concrete prisms to obtain the response of concrete, under non-uniform confinement, to transient loading. Results are compared with the behaviour under monotonic loading. Large strength and ductility enhancement has been obtained in both cases. The cyclically tested specimens displayed a large energy absorption and dissipation capacity, and the method of confinement may be of use in earthquake resistant structures.

INTRODUCTION

Recent tests (1,2) at Imperial College on concrete prisms subjected to a novel form of confinement have shown that considerable gains in strength and ductility over those for plain concrete are possible with this form of reinforcement. The reinforcement, described below, introduces a highly non-uniform and discretised confinement in two orthogonal directions while load is monotonically applied in the third direction. A state of triaxial compression thus develops in the concrete, the benefits of which are well known. These benefits can be important in improving the response, and in particular the ultimate resistance, of structures to seismic excitation. Indeed, in the past, concrete confinement has proved to be crucial to the ability of some structures to avoid collapse during a strong earthquake (3). Some suggestions have been mooted regarding the use of this type of discretised confinement in the construction, strengthening and repair of critical zones of seismic resistant structures (1,2). The large area under the typical stress-strain curves obtained from monotonic loading referred to above suggests that the specimens possess a substantial energy absorption capacity. More pertinent to seismic design, however, is the portion of absorbed energy that will be dissipated in the specimens.

How the same specimens behave under compressive cyclic loading is the subject of this paper. The prisms were subjected to slow cyclic static loading; this work is being extended to specimens of a larger size, subjected to more realistic loading regimes on the shake table at Imperial College.

SPECIMEN DETAILS

All specimens were cast in the upright position, with dimensions of 138 x 138 x 414mm. A rectangular grid of ducts was formed in each prism during casting (Fig. 1); details of the method employed are published elsewhere(1). Into these ducts were inserted 7.8mm diameter high tensile steel bolts which were finger-tightened to provide passive confinement to the concrete. The force from each bolt was distributed to the concrete by mild steel washers, 2.05mm thick and 22.30 mm external diameter.

In preliminary tests, the annular space between each duct and bolt was found to affect specimen behaviour significantly. Therefore, in half the specimens, this space was grouted with an epoxy resin, injection of the resin being undertaken three days before testing.

Figure 1   Specimen details   Figure 2   Loading and instrumentation details

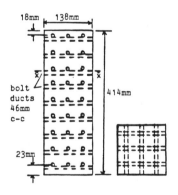
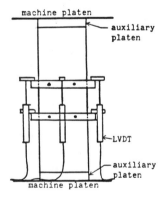

A mix with a characteristic strength of 50 N/mm^2 and maximum aggregate size of 10mm was used. Specimens were cured in water for seven days and stored until testing at a temperature of 20°C and RH of 65%. Control plain prisms of similar size, and standard 100mm cubes were also cast and match-cured. The types of specimens reported in this paper are designated as follows :

    P - plain prisms
    B - bolted prisms, not grouted
    BG - bolted prisms, grouted

LOADING DETAILS

Specimens were concentrically loaded in an Amsler 3000kN testing machine in the (longitudinal) direction perpendicular to the orthogonal arrangement of bolts (Fig. 2). Loading excursions were always compressive, and controlled to give a mean longitudinal strain rate of ±7 microstrains ($\mu\epsilon$)/second. Two regimes of loading were applied :

  a. Simple cycling (Type C1) to maximum load, with the maximum load of each cycle exceeding or matching the maximum load reached in the previous cycle. Unloading was continued until the mean longitudinal stress had been reduced to about 1 N/mm^2. The various load amplitudes of cycling for each specimen type were selected both to highlight the difference between pre-cracking and post-cracking behaviour, and to determine behaviour at high loads.

  b. A series of load cycles which model the amplitudes of typical earthquake response spectra (type C2); this is shown in Fig. 3. Specimens were cycled above a datum of about 0.18 x maximum load attained in monotonic tests.

Figure 3 Type C2 loading regime

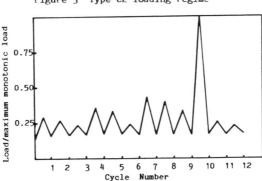

INSTRUMENTATION

A HP3054A data logging system was used to read and process load-deformation data. The applied load was measured by a pressure transducer; the gross prism cross-sectional area was used in the conversion to longitudinal stress σ_1. Four LVDTs mounted between two compressometer frames (Fig. 2) recorded the longitudinal deformations of the four sides of the central third portion of the prism. These were converted to strains and averaged to give the mean longitudinal strain ϵ_1.

EXPERIMENTAL RESULTS

The maximum longitudinal stress attained, σ_m, and the mean longitudinal strain, ϵ_m, at which this occurred in the various specimens are tabulated in Table 1. The table also contains the corresponding values for similar specimen types which had previously been loaded monotonally to σ_m. The behaviour of each test specimen is represented in the graph of σ_1 against ϵ_1 (Fig. 4). The σ_1-ϵ_1 curves for the cyclic tests are also compared with the corresponding monotonic loading curves in this figure. From the initial portions of the reloading curves in Fig. 5 can be seen the degradation during cycling of the longitudinal stiffness of specimens at the start of successive reloadings.

Table 1 Maximum stress σ_m and corresponding mean longitudinal strain ϵ_m attained by specimens

Loading	Specimen	σ_m (N/mm^2)	ϵ_m ($\mu\epsilon$)
C1	1/B	52.1	15450
	1/BG	86.6	26996
C2	2/B	51.7	39687
	2/BG	84.7	21791
Monotonic	3/B	51.3	42628
	3/BG	83.7	20588
	3/P	45.8	1998
	Mean cube strength 70.9 N/mm^2		

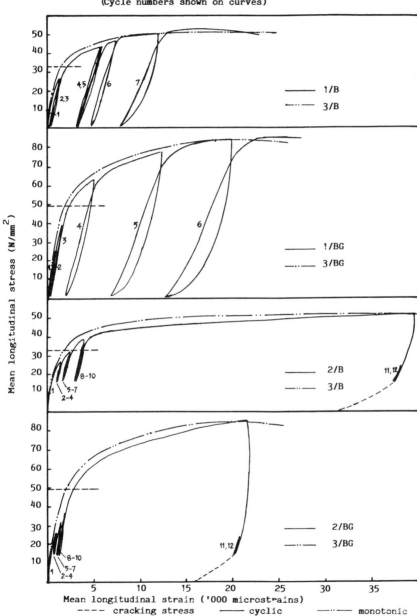

Figure 4 Longitudinal stress-strain curves for all specimens. (Cycle numbers shown on curves)

Figure 5 Initial portions of loading and reloading curves for cyclic test specimens

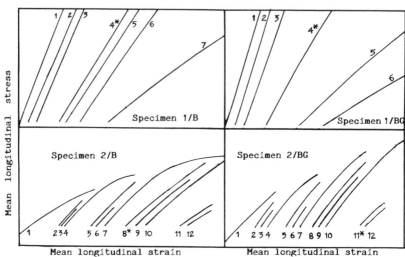

Note : Residual strain for each cycle not drawn to scale.
Numbers refer to sequence of cyclic loading or re-loading.
* refers to first reloading curve after cracking load was reached.

Figure 6 Untested plain prism and specimen 1/B after testing.

Figure 7 Components of dissipated energy (6)

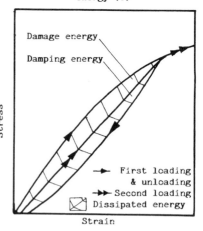

DISCUSSION

Earlier monotonic tests (1,2) have shown that considerable strength or ductility enhancement over plain concrete can be gained by the laterally bolted prisms, as evident in Table 1 and Fig. 4. The grouted specimens were quite distinct from their ungrouted counterparts, showing much higher strength gain while the ungrouted prisms underwent much higher longitudinal deformation near and at maximum load. A feature of the latter was their ability to continue carrying more than 90% of maximum load for large strains beyond ϵ_m. Both types suffered extensive surface spalling and internal damage, as well as significant axial and flexural bolt deformation, at high loads; much energy was apparently expended in producing this large amount of damage, as is also evident from the large area under the monotonic σ_1-ϵ_1 curves. The extensive damage and the longitudinal shortening (which was largely permanent) are seen in Fig. 6. Similar damage was observed in cyclically loaded specimens at the end of the tests. Some features of specimen response under cyclic loading are discussed below.

Envelope curve

To each of the cyclic σ_1-ϵ_1 curves can be drawn an envelope curve (Fig. 4). When this is compared to the corresponding monotonic curve, it can be seen that they are essentially the same. Similar observations have been made for plain concrete (4) and spirally reinforced concrete (5) under cyclic compression. The concept of an envelope curve thus appears to be valid for such reinforced specimens under cyclic loads, and this observation adds to the limited evidence that envelope curves are not only applicable to plain concrete but to reinforced concrete as well.

σ_m and ϵ_m

Table 1 suggests that the maximum stress attained by each specimen type is unaffected by the regime of loading. However, the value of ϵ_m varies considerably. An earlier paper (2) drew attention to the greater inconsistency of the strain values for ungrouted specimens, and attributed this to the weakness due to the presence of the annular ducts. The large variability of strain values was not encountered in monotonic tests of grouted prisms. In the present tests, it appears that the grouted prisms tended to attain higher values of ϵ_m during cycling compared to those attained during monotonic loading. This enhanced strain behaviour may be due to incremental damage and creep deformation under the slow strain rate of cycling.

Stiffness degradation

From Fig. 5, it can be seen that initial longitudinal stiffness is influenced by the load amplitude of cycling. At loads below the cracking load, stiffness degradation was insignificant; once the cracking load was exceeded, stiffness showed a marked deterioration; the higher the load amplitude, the larger the stiffness degradation.

The load amplitude also affected the amount of 'residual' longitudinal strain at the end of each unloading cycle. The residual strain was increased each time a load cycle was taken beyond the highest maximum load reached in any previous cycle. Cycling to less than this maximum load did not produce any additional residual strain, as shown in Fig. 4. Once again, the cracking load marked a change in behaviour - if cracking load was not exceeded, the residual strain was small; once exceeded, the residual strain increased dramatically with each cycle of increasing load amplitude.

Energy dissipation

While it was observed from monotonic tests that the bolted prisms, particularly the ungrouted types, exhibited a high energy absorption capacity, the cyclic tests demonstrated that the absorbed energy was largely dissipated. The amount of dissipated energy was relatively insignificant before the cracking load was reached, and increased greatly once it was exceeded. Some authors (6) have suggested that the dissipated energy is composed of two parts - energy due to 'damage' and that due to some form of damping (Fig. 7). Assuming a similar decomposition for the bolted prisms, it can be inferred from Fig. 4 that much of the dissipated energy was work done in damaging the material tested. It is also apparent that significant damage occurred only during the portions of the load cycles over which the specimen was taken for the first time, often referred to as virgin loading. This was deduced from the observation that repeating a load cycle to the same load produced approximately coincident loading and unloading $\sigma_1-\epsilon_1$ curves, and did not result in any additional residual strain. On the other hand, above the cracking load, virgin loading was always accompanied by increased strain, increased absorbed, dissipated and damaged energy, and increased residual strain.

CONCLUDING REMARKS

The tests reported here are part of a larger series of experiments on discretely confined concrete. In some specimens, the bolts were prestressed to produce active confinement. Loadings also included patch and strip loads, and tests are planned to subject specimens to simulated seismic loading in the Imperial College shake-table facility. The method of confinement may be of application in seismic-resistant structures, for repair and strengthening, and in localised zones of high bearing stresses or load concentrations such as hinges and anchorages.

REFERENCES

(1) PERRY, S H, CHEONG, H K, ARMSTRONG, W E, 'Improved strength and ductility in concrete elements for earthquake resistance', Proceedings, Fourth Canadian Conference on Earthquake Engineering' Vancouver (June 1983).

(2) CHEONG, H K, PERRY, S H, 'Experimental behaviour of concrete elements under non-uniform confinement', Proceedings, International Conference on Concrete under Multiaxial Conditions, Toulouse (May 1984).

(3) GREEN, N B, 'Some structural effects of the San Fernando earthquake', ACI Journal, Vol. 69, No. 5 (May 1972).

(4) KARSAN, I D, JIRSA, J O, 'Behaviour of concrete under compressive loading', Journal of Structural Division, ASCE, Vol. 95, No. ST12 (Dec. 1969).

(5) SHAH, S P, FAFITIS, A, ARNOLD, R, 'Cyclic loading of spirally reinforced concrete', Journal of Structural Engineering, ASCE, Vol. 109, No. 7 (Jul. 1983).

(6) SPOONER, D C, DOUGILL, J W, 'A quantitative assessment of damage sustained in concrete during compressive loading', Magazine of Concrete Research, Vol. 27, No. 92 (Sep. 1975).

DISCUSSION

Dr C D Goode   University of Manchester

Table 1 indicates that the bolted and grouted specimens showed considerable increase in strength over the ungrouted specimens and over the plain prisms illustrating the enhancement achieved by triaxial compression. It would however seem unfair to compare the bolted specimens which were ungrouted with the solid plain prisms (even though a slight increase in strength took place) because the holes would cause weakness. Comparison with plain prisms with holes would be preferable and I would like to ask the authors, if such tests have been carried out, what their strength was.

The stress-strain curves in Fig 4 became concave to the stress axis at the later load cycles. I have observed a similar behaviour with concrete which has been subjected to fatigue loading in compression and would be interested to hear the authors' views on the reasons for this 'stiffening' as the stress increases.

Author's Reply.

We did test a few prisms which were provided with the ducts but which were not reinforced at all. Expectedly, the strength was very low, about 25 MPa. If this value is compared with the strength of a corresponding bolted specimen (3/B) of 51.3 MPa, one can see that there is a strength enhancement of about 100%.

I am glad that Dr Goode has drawn attention to the concave feature in the initial portions of some of the reloading curves. This seemed to occur after the prisms had been cycled to high loads, and had undergone significant longitudinal strain and disruption. We believe that this feature resulted from the re-opening, during re-loading, of those cracks which had completely or partially closed up during the previous unloading. As re-loading proceeded, the re-opening of the cracks was impeded by an increasing confinement from the bolts, with a resulting 'stiffening' of the prisms.

7 Production and use of high yield model reinforcement

F. A. NOOR, S. RAVEENDRAN and W. EVANS
North East London Polytechnic, UK

SUMMARY

A method of producing deformed model reinforcement in the range of 2.0 to 4.0 mm, effective diameter, is described. The new material has both the strength and yield characteristics similar to those of prototype hot rolled high yield steel bars. The methods of fabrication and testing of small scale models using the new reinforcement and glass bead microconcrete are outlined. The results obtained from a recent study, using these techniques, are of interest to both research and design engineers.

INTRODUCTION

A number of research reports [1-3], published over the past fifteen years, have shown the need for deformed model reinforcement for microconcrete models. Commercially available threaded rod has been one of the materials considered.[3] Unfortunately, this cannot be used to model both the bond and effective diameter at the same time. The method of fabricating deformed mild steel wire, described earlier[2], makes it possible to vary both the spacing and height of ribs, in order to simulate the bond of prototype steel bars with various surface characteristics. However, mild steel shows a considerable reduction in ductility if it is not sufficiently annealed after cold rolling. This may occur if the cold worked material is heat-treated, just sufficiently, to obtain a yield strength above 400 N/mm². The resulting partial recrystallisation of the ferrite grains gives an unacceptable variation in its strength. These disadvantages also apply to the threaded rod and other types of cold worked mild steel model reinforcement. There appeared to be an urgent need for a material, the stress-strain relationship of which exhibited a reasonably sharp yield point, an adequate length of yield plateau and elongation at failure. It should be noted that the presence of a long yield plateau would be helpful in tests of secondary effects, such as membrane action in slabs. The other considerations are that the variation of yield strength, within a batch, in the range of 450-600 N/mm², should be small and it should be possible to produce straight bars in effective diameters ranging from 2 to 4 mm. Such a model reinforcement has been developed and its method of production is described.

Research on modelling of concrete structures, at NELP, over the past five years, has led to a number of new techniques. The application of some of these methods will also be outlined.

MEDIUM CARBON STEEL

A possible approach to the problem of modelling high strength reinforcement could have been to obtain the material normally used in hot rolled high yield prototype bars. It was considered unlikely that this would be produced in wire dimensions. In any case, its high strength is given by the control of ferrite grain size and micro-alloying, in a low carbon steel. Any cold working and subsequent reheating for recrystallisation would lose its strength mechanism and the desirable properties would not be obtained. The only way forward was to try a medium carbon wire suggested earlier[2], and at first, a material with a carbon content of 0.6% was heat-treated at 650°C before and after cold rolling. This proved to be too hard and neither the strength nor the required surface deformations were obtained. The subsequent trials were conducted on a material with a little lower carbon content and these proved to be more successful. The periods of heat treatment, before and after cold rolling, varied with the carbon content, the diameter of the wire, the degree of surface deformations and the desired strength.

Typical stress-strain relationships of 2.4 mm nominal diameter reinforcement, for different yield strengths, are shown in Fig.1. It can be seen that in a model reinforcement with a yield strength, f_y, in the range of 475 to 650 N/mm², there is a sharp yield point, a yield plateau extending beyond a strain, ε_o, of 0.035, a strain hardening effect and an ultimate strain, ε_u, greater than 0.1. These properties are close to those of hot rolled high yield prototype bars.

Rolling Process

The machine described in an earlier report[2], was slightly modified to deform the medium carbon steel wire but a number of problems remained. It was then decided to develop a new rolling mill so that sufficient quantities of the reinforcement could be produced for several research programmes. The possibility of supplying this material to other research laboratories was not overlooked and it is hoped to do this in the near future. Many of the disadvantages, present in the original wire rolling machine, have been eliminated by the adaptation of a larger and more sophisticated Hille 25 strip rolling mill. The new machine is capable of producing a model reinforcement with effective diameters close to 2, 2.4, 3.2 and 4.0 mm, respectively. Fig.2 shows both the cross-section and the pattern of ribs of the model steel. The sides of these ribs are vertical and no attempt has been made to match their profile with that of prototype bars. The minimum width is that dictated by the cutting tools and research carried out by others[4] shows that the influence of rib profile is not significant. The effective diameter, D_e, depends on the relative height of the rib and is given:

$$D_e = KD_1 D_c \qquad \text{............} \quad (1)$$

The constant K could be determined by comparing the mass of a metre length of plain and deformed wire and gives a rapid evaluation of the effective diameter.

The initial heat treatment may be carried out in a commercial vacuum furnace when the wire is in a coil form. Alternatively, the wire may be cut in about one metre lengths and treated in a specially designed nitrogen atmosphere tube furnace. The heat-treatment after rolling can, in some cases, be as long as 16 hours. The tube furnace is then more economical and would normally be used.

Tensile Testing

The most efficient method available was to use a 500 kN Dartec Servo-Hydraulic Machine with a 50 kN load cell. A specially designed wire extensometer gave accurate charts of stress-strain relationship, such as those shown in Fig.1. In developing the new reinforcement a large number of tensile tests have been carried out. These have shown that the variation in strength, between batches and along the length of wires, may easily be kept below ±2%.

USE OF HIGH YIELD MODEL REINFORCEMENT

Recent research[5] has shown that for a given compressive strength, the tensile strength and the stress-strain relationships of microconcrete may be varied by the type of aggregate used. Whilst plain model reinforcement is still used by some researchers, recent reports[3,5] leave no doubt that, for accurate similitude of crack spacing and width, the ratio of rib height to effective diameter is very important. Our current practice is to use a ratio of 0.06 which not only gives sufficient anchorage but also good cracking similitude.

Effect of Shear on Deflection of Reinforced Concrete Beams

A number of continuous beams tested during the past three years have shown the experimental deflections to be significantly greater than those given by theoretical calculations, based on flexural deformations. This was observed both in reinforced concrete beams of 100x150 mm in cross-section and their one fifth scale microconcrete models. The influence of shear on deflections was suspected and further evidence of this was provided by tests reported by Monnier[6]. The authors' view was that it is not difficult to model the moment-curvature relationship and deflections of beams if the correct percentage of deformed model reinforcement is used and a series of model tests were planned. Table 1 gives the experimental programme for different types of beams. The object of the simply supported tests was to obtain the moment-curvature relationships in the constant moment zone and typical curves are given in Fig.3. This data was then used to predict the deflection of all beams.

Table 1   Beam Tests

Type of Test	Beam No	Main Rein't	Span mm	f_y N/mm²	DEFLECTION Δ_E mm	DEFLECTION $\dfrac{(\delta_s + \delta_{cs})}{\delta_f}$	DEFLECTION Δ_E/D_T
Simply supported	SL1a SL1b	2Y1.95	400	472 466	1.65 1.70	5.5 5.5	1.00 1.02
	SH1a SH1b	2Y2.84	400	510 513	2.79 2.81	5.3 5.3	1.02 1.03
Cantilevered	CAL1a CAL1b	2Y1.95	100	472 469	0.38 0.40	33.1 33.1	1.03 1.08
	CAL2a CAL2b	2Y1.95	300	482 477	2.80 2.82	6.3 6.3	1.06 1.07
	CAL3a CAL3b	2Y1.95	500	477 474	7.92 7.80	3.2 3.2	1.12 1.10
	CAH1a CAH1b	2Y2.84	100	513 513	0.61 0.60	30.3 30.3	1.05 1.03
	CAH2a CAH2b	2Y2.84	300	516 513	4.59 4.54	4.8 4.8	1.06 1.05
	CAH3a CAH3b	2Y2.84	500	513 519	12.05 12.28	2.2 2.2	1.02 1.04
Continuous	COL1a COL1b	2Y1.95	300	467 469	0.63 0.64	38.2 38.2	1.14 1.15
	COH1a COH1b	2Y2.84	300	518 510	1.07 1.06	32.2 32.2	1.14 1.13

Δ_E, Δ_T = Experimental and theoretical deflections, respectively (shear effects included)

$\delta_s, \delta_{cs}, \delta_f$ = Theoretical deflection due to shear, increase in curvature in the presence of shear and bending, respectively.

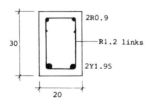

Experimental Details

The method of fixing the reinforcement was to use 0.3 mm diameter constanton wire and this gave fairly rigid cages. The microconcrete mix used 2 mm glass beads and details of this are given in another report[5].

The beams were tested under displacement control and the loading arrangements for the simply supported beams are shown in Fig.4. The continuous beams incorporated a load cell at the central support and thus an accurate assessment of the experimental moments could be made.

Theoretical Calculations

The results of deflections calculated at 0.6 of theoretical yield load are also given in Table 1. The expressions used to calculate the various theoretical values are given elsewhere[6]. An allowance was made for variation in flexural stiffness along the beam by dividing the beam into segments and determining the curvatures due to bending from the experimental data. Using the principle of virtual work, the product $M_1 M_o/EI$ was integrated over the whole length of the beam using Simpson's Rule whilst $M_1 \phi_o$ was integrated only over the cracked length. The deformations in the uncracked part were considered to be insignificant both in the case of shear strain, as well as the increase in curvature in the presence of shear, ϕ_o, and have been neglected.

Results

The results show that the use of experimental curvatures due to bending leads to an accurate prediction of midspan deflections of simply supported beams. The deflections due to shear are only a small fraction of the total displacements and may be ignored. In the case of short span cantilevered and continuous beams, the error in the theoretical values may be as high as 38%, if shear deformations are ignored. The reason for the reduction in error with the highly reinforced beam is that the increase in curvature in the presence of shear, ϕ_o, is inversely proportional to the steel percentage.

CONCLUSIONS

1. It is possible to obtain the properties of hot rolled high yield steel reinforcing bars, used in structural concrete, by cold rolling ribs into a medium carbon steel wire. The variation in yield strength, within the range of 450 to 600 N/mm², can be maintained below ±2%.

2. The use of new material, in model beams made from glass bead microconcrete, gives a realistic pattern of cracking and is helpful in teaching, research and design of concrete structures.

REFERENCES

(1) HARRIS, H G, SABNIS, G M and WHITE, R N, 'Reinforcement for small scale direct models of concrete structures' American Concrete Institute (1970) ACI Publication No.SP-24.

(2) NOOR, F A and KHALID, M, 'Deformed wire reinforcement for microconcrete models', Reinforced and Prestressed Microconcrete Models. Construction Press (1980) pp.103-118

(3) EVANS, D J and CLARKE, J L, 'A comparison between the flexural behaviour of small scale microconcrete beams and that of prototype beams', London, Cement and Concrete Association (March 1981) Technical Report 542.

(4) SKOROBOGATOV, S M and EDWARDS, A D, 'The influence of the geometry of deformed steel bars on their bond strength in concrete' Proceedings of the Institution of Civil Engineers, Part 2, Vol.67 (June 1979)

(5) NOOR, F A and WIJAYASRI, S, 'Modelling the stress-strain relationship of structural concrete'. Magazine of Concrete Research Vol.34, No.118 (March 1982).

(6) MONNIER, I T, 'The behaviour of continuous beams in reinforced concrete'. Research Report BI-69-71. Institute TNO for Building Materials and Building Structures, Netherlands (September 1969)

(7) UJOODHA, N K, 'Effect of shear on deflections of reinforced concrete beams', North East London Polytechnic (1984) Undergraduate Project Report.

ACKNOWLEDGEMENTS

The authors wish to record their gratitude to the Science and Engineering Research Council, which funded the work over the past two years.

Figure 1 Typical stress-strain relationship of 2.4 mm diameter reinforcement

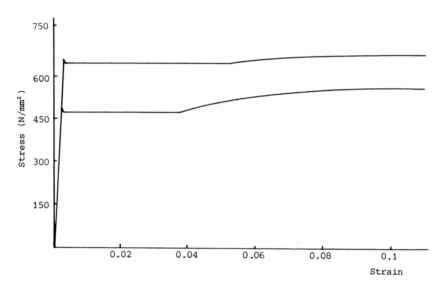

Figure 2 Typical cross-section and pattern of ribs of model reinforcement

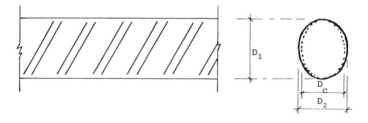

Figure 3 Moment-curvature relationship of simply supported beams in the constant moment zone

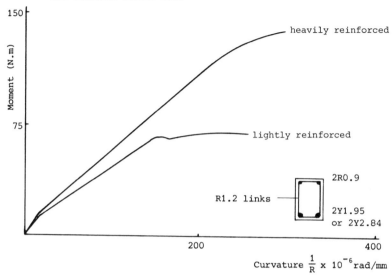

Figure 4 Rig for testing model beams

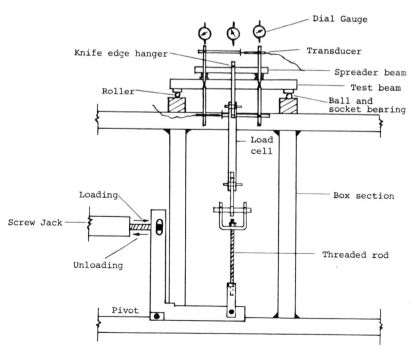

DISCUSSION

F M Falih   University of Sheffield   UK

How are the reinforcement cages for the small-scale model beams fixed accurately.

Author's Reply.

The 0.3 mm Constanton wire, mentioned in the paper, was used in the same way as the larger diameter mild steel wire, when used to fabricate prototype reinforcement cages. The accuracy of links is important and these were made by winding an appropriate material around a steel form, with a cross-section similar to that of links. The rectangular helix formed was cut at appropriate places to obtain either single or double links.

8 Adaptive control technique for the simulation of seismic actions

J. M. JERVIS PEREIRA and F. J. CARVALHAL
Laboratório Nacional de Engenharia Civil, Portugal

Summary: The paper describes the main aspects of an experimental technique, using adaptive control, to subject structure models to excitations, with prescribed stochastic characteristics, simulating seismic actions. The control algorithms for two alternative cases, self-tuning of power spectral density or of response spectrum, are presented as well as a general description of the main experimental rig and typical obtained results.

1 - INTRODUCTION

Several studies carried out on the seismicity of the portuguese territory, lead to the establishement of seismic risk charts and to the characterization of seismic actions for epicentres located at either near or far distances. In Portugal, the new and recently approved Code for Safety and Actions for Buildings and Bridges Structures, contains a relevant part regarding the definition of seismic actions based on the acceleration power spectral density (PSD) or on the acceleration response spectrum (RS). The LNEC has developed an experimental technique which hinges on the use of an adaptive (or self-tuning) control scheme, aining to ensure, within certain limits, that the structure model (or small prototype) when mounted on a shaking table is subjected to excitations with prescribed stochastic characteristics The control scheme generates a zero mean stationary gaussian excitation with an adequate power spectral density or, alternatively, a response spectrum. It works on an iterative fashion, automatically self-tuning to the dynamics of the model to be tested. Within fair limits, it can also accomodate to reasonable and unpredictable changes of the model dynamic characteristics . The main purpose of the paper is simply to outline the main conceptual

aspects of the above mentioned technique, under the viewpoint of its use on
model analysis, without going thus into too much theoretical or even
practical details which could be distracting. Taking this into consideration
the paper briefly refers to the basic theoretical concepts of the control
algorithms for the two cases, one concerned with the power spectral density
and the other with the response spectrum, and presents a general description
of main experimental rig where the methods is implemented and which is now
being used. Also, some results obtained an this rig are presented.

2 - BASIC CONCEPTS OF THE METHOD

2.1 - General considerations

Rather basically, an experimental rig (or testing system) for unidirectional
seismic tests consists of an actuator system, usually a position
servosystems, which drives a table upon which the model is mounted. For
simultaneous multidirectional tests, the complexity naturally increases but
the basic configuration remains similar. We shall restrict ourselves, however,
to the unidirectional tests. The testing system input is a signal which is
fed into the actuator (e.g. a voltage signal) and the most common output is
the table response represented by a variable associated with its movement
(eg. table acceleration). There exists a dynamical relationship, widely
referred to as transfer function, between the output and the input. The main
task of the technique is thus, given this dynamical relationship, to generate
an adequate input in order to have an output such that the table response
has the required stochastic characteristics, regardless of the model being
tested. An important point to focus is that the model characteristics heavily
affect the aforesaid dynamical relationship; this is the reason why we have
settled for an adaptive control scheme which senses that relationship by
means of an implicit non-parametric identification of its characteristics .
Most of the work concerned with this particular technique has been carried
out from 1974 to 1979 (1) (2) (3); a very comprehensive paper (4) on closily
related subjects, namely the use of non-parametric algorithms on self-tuning
regulators has been recently published.

2.2 - Control Scheme Basic Concepts

Both the aforesaid versions considered to implement the method, i.e., either
aiming to adapt a PSD or a RS, share some fundamental conceptual aspects.
Actually, they differ in what is concerned with the algorithms of some of
the control scheme main stages. It is to be noted that the selft-tuning

outer loop does not perform a "follower" type of control action. In fact, the control scheme works based not on the testing system output itself, but rather based on a "performance index" (or cost function) evaluated from that output. This "performance index" is simply the PSD or RS of the output. The control scheme works on an iterative manner and is implemented in a digital compute. We have thus a hybrid system where the transition between the discrete and the continuous domains are carried out though digital to analog and analog to digal converters. For description purpose, we way consider that the control action involves, in each cycle of iteration,three main stages, as follows:

1) The performance index, that is to say, the PSD or RS, is estimated from the N-point time series obtained from the continuous output signal through sampling at Δt time intervals, corresponding thus to a "frame" of the output with a time lenght $T = N.\Delta t$.

2) The performance index so obtained is then subjected to an implicit comparison with the Reference, which is the prescribed PSD or RS for the test, by means of an adequate "decision" algorithm. A series of power spectral values are computed through this algorithm.

3) The values computed at the previous stage are used to generate the input time series for the next iteration cycle, which will have the suitable power spectrum and an approximate zero mean gaussian amplitude distribution. This stage may be called the "test signal synthesizer" (4).

The testing system plus model (the controlled system) is assumed to be linear and time-invariant. This assumption, naturally, diverge from the reality . Nevertheless, the characteristics of the controlled system may vary in time provided that this variation is slow in comparison with the output frame time lenght T. Moreover, the self-tuning control scheme can cope with some non-linearities, provided that they will be kept within "reasonable" limits, being regarded then as system perturbations.

2.3 - <u>Self-tuning of Power Spectral Density</u>

In this alternative the performance index is the power spectral density as we have already mentioned. The iteration cycle will be denoted by index i, the discrete frequencies by f_k and the input and output of the controlled system by u and y, respectively.

decision algorithm takes an important an significant advantage of this by generating spectral values $S_u(f_k)$ for each discrete frequency, independently of the others. Smoothing procedures may be used, as an option, to reduce variability of the spectral estimates $S_y(f_k)$. However, in order to preserve the aforesaid orthogonality, it is most convenient not to use the frequency averaging type of smoothing procedures but rather the frame averaging type (5). The discussion in detail of this particular matter (mainly in what regards its significant influence on the convergence of the control scheme) is, however, outside the scope of this paper; it may be found in literature, for instance and among others, in references (3)(4) and (5).

2.4 - Self-tuning of Response Spectrum

The response spectrum (RS) may be referred to as a series of spectral values, each one assigned to a generic frequency f_n, representing the observed maximum of the response (or output) of a one degree-of-freedom linear oscillator, to a certain input excitation y(t), the oscillator being characterized by its natural frequency f_n and its damping factor η. In seismic actions simulation case, the RS is usually an acceleration response spectrum defined at N_f 1/3 of octave normalized frequencies $f_n (n=0,1,...,N_f-1)$. A prescribed common damping factor η is adopted, being then a fixed test parameter. The prescribed or reference RS is denoted by $R_r(f_n)$.

2.4.1 - Algorithm

The control algorithm is also considered with three member stages:estimator, decision and test signal synthesizer. In the estimator stage the Iwan's matrix algorithm is used to estimate be response spectrum, $R_y(f_n)$, of the testing system output time series y(m) (m=0,1,...,N-1) with time length T=NΔt, where Δt denotes the sampling time interval. The syntheziser stage hinges on the same algorithm used in the PSD case (see 2.3.1), the difference being confined to the computation of the spectral values $U(f_k)$ to generate the input time series $u(m)_{i+1}$. The computation of these spectral values $U(f_k)$ is the main task of the decision stage. We think that it is worthy to explain the reasoning behind the decision algorithm with a reasonable degree of detail. Before doing it however, it has to be pointed out that the RS case implies the generation of an initial input time series based on specific spectral values $U(f_k)_o$; we shall describe the procedure for this initial input later on. In a generic cycle i, the system response to an input $u(m)_i$, given by expression (4), may be expressed as

2.3.1 - Algorithm

The controlled system output time series, a sequence of N data values $y(m)$ $(m=0,1,...,N-1)$, is subjected to a Fact Fourier Transform (FFT) algorithm, thus producing the Fourier coefficients $Y(f_k)$ where $f_k = k/N\Delta t$; Δt denotes the sampling time interval, being $T = N.\Delta t$. The raw power spectrum estimates (or periodogram) over a cycle i are then estimated as follows

$$S_y(f_k)_i = \frac{1}{T} |Y(f_k)_i|^2 \qquad (1)$$

The important part is the decision stage algorithm which is given by

$$S_u(f_k)_{i+1} = S_r(f_k) \cdot \frac{S_u(f_k)_i}{S_y(f_k)_i} \qquad (2)$$

where $S_r(f_k)$ denotes the reference power spectral density values, defined at the frequencies f_k. This algorithm assumes linearity and absence of extraneous noise, accepting thus as completely valid the known relationship (5)

$$S_y(f_k)_i = |G(f_k)|^2 \cdot S_u(f_k)_i \qquad (3)$$

where $G(f_k)$ stands for the controlled system transfer function. For each cycle i+1, the test signal synthesizer algorithm is based on values $S_u(f_k)_{i+1}$ computed at the decision stage. It generates a pseudo-random time series $u(m)_{i+1}$, with period T, power spectral density $S_u(f_k)_{i+1}$ and an approximate zero mean gaussian amplitude distribution. It uses, for this purpose, the inverse discrete FFT such that

$$u(m)_{i+1} = \frac{1}{T} \sum_{K=0}^{N-1} U(f_k)_{i+1} \exp\{j2\pi km/N\} \qquad (4)$$

where the Fourier coefficients $U(f_k)_{i+1}$ have modulus equal to $\{T.S_u(f_k)_{i+1}\}^{1/2}$ and phase values $\phi(f_k)$ which are randomized (3)(4)(6). When starting any particular test a pseudo-random phase value is assigned to each $\phi(f_k)$ and, aftewards, maintained thoughout the same test.

2.3.2 - Comments on the Algorithm

We should like to emphasize that this algorithm allows conducting the tests without knowing, in an explicit way, the dynamic characteristics of the model, due to the existence of an implicit non-parametric identification of those dynamics at the decision stage. Taking into account the algorithm structures of the synthesizer and the estimator stages, the spectral estimates $S_y(f_k)$ are orthogonal at different discrete frequencies $f_k(4)$. The

$$y(m)_i = \frac{1}{T} \sum_{K=0}^{N-1} U(f_k) \cdot G(f_k) \cdot \exp\{j\,2\Pi\,km/N\} \qquad (5)$$

which, passing through a procedure to evaluate response spectrum value at frequencies f_n, will give rise to N_f new random series expressed as

$$z_n(m)_i = \frac{1}{T} \sum_{K=0}^{N-1} U(f_k) \cdot G(f_k) \cdot H_n(f_k) \exp\{j2\Pi\,km/N\} \qquad (6)$$

where $H_n(f_k) \equiv H(f_k; f_n, \eta)$ denotes the transfer function of the oscillator with natural frequency f_n an damping factor η; this damping factor is, naturally, equal to the one adopted for the reference RS. This new time series $Z_n(m)$ follows a probabilistic distribution approximately given by Cramér (7)(8) formula; with probability $P(x)$ we shall have (dropping index i)

$$\max_m\,Z_n(m) = \sqrt{2\lambda(f_n)\{\ln f_n T - \ln \ln P(x)^{-1}\}} \qquad (7)$$

where

$$\lambda(f_n) = \frac{1}{T^2} \sum_{K=0}^{N-1} |U(f_k)|^2 \cdot |G(f_k)|^2 \cdot |H_n(f_k)|^2 \qquad (8)$$

At this point we accept the following simplifications:

1) Since the convergence criterion will be referred to the RS values corresponding to the N_f frequencies f_n, the spectral values $U(f_k)$ will be considered as being constant in frequency intervals such that

$$U(f_k) = U(f_n) \quad \text{for } \sqrt{f_{n-1} \cdot f_n} < f_k < \sqrt{f_n \cdot f_{n+1}} \qquad (9)$$

2) Assuming low values for the damping factor, the oscillator transfer function will be considered as follows

$$\begin{aligned}|H_n(f_k)| &= 1 & 0 < f_k &< (1-\eta)\,f_n \\ |H_n(f_k)| &= 1/2\,\eta & (1-\eta)f_n < f_k &< (1+\eta)\,f_n \\ |H_n(f_k)| &= 0 & f_k &> (1+\eta)\,f_n\end{aligned} \qquad (10)$$

Taking these simplifications into consideration, expression (8) may decomposed und, after some mathematical manipulation, we have

$$\lambda(f_n) = (f_n) + |U(f_n)|^2 |G(f_n)|^2 \cdot \frac{f_n}{2\eta} \qquad (11)$$

where the term $\lambda(f_n)$ accounts for the contributions of all the normalized frequencies f_1 with $1<n$; thus, this term is zero for the lowest normalized frequency being considered, i.e., $\lambda(f_o) = 0$. For seismic actions, it is acceptable (9) to take $P(x) = 0.5$ and we may consider that $R_y(f_n)$, the mean

maximum value, is given by expression (7) together with the simplified expression (11). The decision algorithm is then expressed as

$$U(f_n)_{i+1} = r(f_n)_i \cdot U(f_n)_i \quad (12)$$

where

$$r(f_n)_i = \left[\frac{R_x^2(f_n) - 2\lambda(f_n)_i \ln\{f_n T + 0.37\}}{R_y^2(f_n) - 2\lambda(f_n)_i \ln\{f_n T + 0.37\}} \right]^{1/2} \quad (13)$$

The spectral values $U(f_k)_{i+1}$, for $k \neq n$, being computed by means of expression (9). In this algorithm the decision factor $r(f_n)$ is actually bounded in the range from 0.6 to 1.7, due to numerical stability problems which may arise mainly in the first iteration cycles. We shall now examine the generation of the initial input time series. To begin with, when starting a particular test a specific cycle is carried out, its purpose being to estimate, through a non-parametric identification procedure, the values of $|G(f_n)|^2$ used to compute $\lambda(f_n)_i$ as well as $U(f_n)_o$. For a zero damping linear oscillator, with natural frequency f_n, subjected to an acceleration input $y(t)$, the corresponding velocity RS is such that $R_v(f_{n,0}) > |Y(f_n)|$. On the other hand, it is almost evident that when synthesizing the initial input the aim is to obtain $R_y(f_n)_o \simeq R_r(f_n)$. Use may then compute the spectral values for the initial input time series through

$$U(f_n)_o = \alpha \frac{R_r(f_n) \beta(f_n)}{2\Pi f_n |G(f_n)|} \quad (14)$$

together with expression (9), where $\beta(f_n) = \{1+0.5\eta T\, 2\Pi f_n\}^{1/2}$ is a correction factor due to the non-zero damping factor and $\alpha < 1$ (usually in the range from 0.5 to 1).

2.4.2 - Comments on the Algorithm

We note that in this control algorithm a previous cycle is need to estimate $|G(f_n)|$ and also that these values are not updated is the following iteration cycles the implementation of a procedure tu update them would not be difficult but would imply a significant increase in computational effort. Since the RS case does not have an orthogonality property as the one referred to in section 2.3 for the PSD case, the decision algorithm aims to isolate each frequency f_n from the contributions of the other ones. The two major simplifications introduced through expression (8) and (9) are considered to be quite acceptable and are intended to decrease the decision algorithm complexity and the corresponding computational effort.

3 - EXPERIMENTAL RIG

3.1 - Description and Characteristics

The main experimental rig now being used at LNEC consists of two tables, one for horizontal displacements the other for vertical displacements. The horizontal one has effective dimensions of 3x2 makes, maximum load capacity of 70 KN and weighs 30 KN. The vertical one has 2.45x1.45 makes, maximum load capacity of 50 KN and weighs 12 KN. Both tables are drivenn not simultaneously, by a position servohydraulic system. The hydraulic cylinder (actuator) has a stooke of 0.2 metres and can deliver a maximum dynamic force of 200 KN with an oil suplly line pressure of 20 MPa. The hydraulic power supply has a maximum flow rate of 240 litres/min, with a 110 KN pumpmotor. The self-tuning control scheme algorithms are implemented in a central digital computer, the communications between the rig and the computer being performed by a dedicate microprocessor-based system.

3.2 - Obtained Results

In Fig. 1 and Fig. 2, two of the typical results obtained using the technique and the above mentioned rig, are shown. Fig. 1 refers to the PSD case, dashed line representing the reference and full line the system output. Fig. 2 refers to the RS case, dotted line representing the reference and dashed line the system output. In both cases, the aim has been to would be greater than the references ones, in order to couply with seismic Code rules.

Figure 1 - Self tuning of PSD   Figure 2 - Self tuning of RS

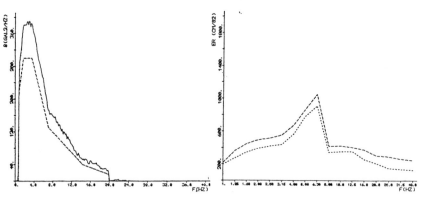

REFERENCES

(1) CARVALHAL, F J, Simulation of Seismic Disturbances Using Adaptive Control, MSc Dissertation, UMIST, Manchester, (1974).

(2) WELLSTEAD, PE e CARVALHAL, F J, 'Digital computer simulation of seismic disturbances', Proc. UKS6 Conference on Computer Simulation, Bowness, Windermere, (May 1975).

(3) CARVALHAL, F J, Control Adaptativo na Simulação de Solicitações Sísmicas Thesis, LNEC, Lisbon, (January 1979).

(4) WELLSTEAD, PE & ZARROP, M B, 'Self-tuning: non-parametric algorithms', International Journal of Control, Vol 37, No. 4, (1983).

(5) BENDAT, J S & PIERSOL, AG, Random Data: Analysis and Measurements Procedures, John Wiley & Sons, (1971).

(6) WELLSTEAD, PE, "Pseudo-noise test signals and the fast fourier transform", Electronic Letters, Vol. 11, No. 10, (May 1975).

(7) CRAMÉR & LEABEETTER, Stationary and Related Stochastic Processes, John Wiley & Sons, (1967).

(8) JERVIS PEREIRA, J M, Métodos Probabilísticos em Engenharia Sísmica, Senior Research Officer Thesis, LNEC, Lisbon, (1974).

(9) FERRY BORGES, J. 'Statistical estimate of seismic loading', International Association for Bridge and Structural Engineering, V Congress, (1956).

Section 1   Papers 1-8

Commentary by W. GENE CORLEY
(Executive Director, Engineering and Resource Development, Construction Technology Laboratories, Illinois, USA)

INTRODUCTION

Because of physical and financial limitations on tests of full-scale structures, models have historically been used to extrapolate design procedures. Modelling techniques and rules of similitude have been well developed over the last century. Nonetheless, the papers reported in this section show that further refinements are still being made.

MICROCONCRETE FOR STRUCTURAL MODEL ANALYSIS

In the first paper of this section, Dr Muller reports meticulous modelling of microconcrete for use in structural analysis. Through careful proportioning of aggregates, microconcretes having appropriate compressive strengths are developed. A unique coating technique is used to reduce tensile strength of the model materials to obtain a suitable relationship with compressive strength. As the author notes, the resulting microconcretes provide good similitude with prototype concretes.

Dr Muller reviewed procedures for preparing modelling reinforcement to obtain suitable bond and tensile properties. Appropriate combinations of annealing, cold working, and knurling are described. These procedures produced bond characteristics and tensile properties similar to prototype reinforcement.

The paper notes that final results give performance in his small specimens that compares well with prototype test data reported in the literature. Results for tests of specimens ranging from 1:11 to 1:5 scale showed good agreement.

In a final comparison, the author suggests that shear tests of slabs can also be suitably modelled. However, care should be taken in trying to extrapolate shear strengths of small-scale models directly to full-scale structures. As indicated by many tests reported in the literature, suitable rules for similitude of shear strength in slabs are not yet fully developed.

USING MODELS FOR STUDYING REINFORCED FIBRE CONCRETE BEHAVIOUR

Craig, Loy and Ryden have reported an ambitious attempt to use microconcrete and gypsum models with fibre to mimic behaviour of fibre reinforced concrete. Techniques are described for fabricating small specimens rapidly.

The paper reports several difficulties in obtaining similitude. In particular, somewhat distorted fibres were used in the model to obtain results comparable to those of the prototype. The authors interpret the findings to imply that the scale of models containing fibre should be kept at 1:4 and larger.

Some question can be raised about the desirability of the authors' use of staples in the models to represent fibres in the prototype. Further study is needed to see if such a distortion is reasonable.

DEVELOPMENT OF A SMALL AGGREGATE CONCRETE FOR STRUCTURAL SIMILITDUE OF SLAB-COLUMN CONNECTIONS

In this paper, the authors provide significant insight into similitude of slab to column connections. Microconcrete mixes using large amounts of fly-ash were developed. Superplasticisers were used to obtain workability. Mixes were developed that gave tensile strength to compressive strength ratios similar to those of prototype concrete.

Reinforcement used in the models was either as received plain mild steel wire or cold drawn knurled wire.

Results of tests on 1:3, 1:4, and 1:6 scale models were reported. Findings of the report indicate that scaling has a significant influence on both flexural cracking and shear strength. However, little effect was found on flexural strength. The authors conclude that punching shear failures of models smaller than about one-third scale significantly over-estimate capacity of the prototype.

THE USE OF PUSHOUT TEST TO SIMULATE SHEAR STUD CONNECTORS IN COMPOSITE BEAM CONSTRUCTION

Tests reported in this paper are used to evaluate the proposed BS 5950 design method for composite beams using through deck welded studs and profiled steel sheets. Push-out tests were carried out and results were compared with tests of full-scale composite beams.

The authors found that the "push-out" test recommended in BS 5950 significantly under-estimates strength and stiffness. A new push-out test that better represents behaviour of the prototype is suggested. Details of the tests are described in the paper. Alternatively, the authors suggest that a factor of 1.8/1.4 be applied to standard tests to recognise higher strengths and stiffness of the prototype.

SMALL SCALE MODELLING TECHNIQUES FOR INVESTIGATING THE FAILURE OF A BIN STRUCTURE

White describes a unique model to determine behaviour of a steel bin that failed in service. The 1:12 scale test model was constructed using rigid vinyl sheeting and polyvinylchloride columns. White noted that vinyls offer advantages over Plexiglas because of more uniform thickness and lower creep.

In addition to describing the modelling techniques for the bin material, a highly unusual combination of loading materials was described. Pre-cooked dry rice and rice plus cracked corn were found to give suitable densities and angles of repose. Although the author reported significant local buckling occurred in the model, it was not noted whether the observed behaviour of the model suitably represented the observed failure of the prototype.

TRANSIENT LOADING OF STRUCTURAL CONCRETE

A unique system using discrete confinement of concrete was evaluated and reported in this paper. Confinement was provided by rows of bolts running orthogonally through concrete prisms. In some cases, the bolts were pre-tensioned to obtain active confinement.

Results suggest that even when cyclic loading is applied, extremely large strain capacity can be developed through use of confinement. Strains in the range of 2 to 4% were obtained.

The authors suggest that discrete confinement might be suitable for retrofitting damaged and existing under-designed structures. No information is given on similarity between results obtained with the relatively small tests reported and those of prototype structures. Information on similitude seems necessary before widespread use of the procedure can be considered.

PRODUCTION AND USE OF HIGH YIELD MODEL REINFORCEMENT

In this paper, Noor, Raveendran and Evans describe methods for producing deformed model reinforcement having diameters ranging from 2 to 4 mm. They report that the reinforcement has strength and yield characteristics similar to hot rolled high yield steel bars. Results of use of these bars in small models using glass bead microconcrete are presented.

In this paper, deformations on wire are obtained using a rolling device rather than a knurling device. Both height and spacing of deformations can be varied to give desired bond characteristics.

Although yield stresses and strengths of the reinforcement tend to be similar of those for hot rolled high strength bars, the yield plateau tends to be slightly on the high side. In general, however, results are extremely promising.

ADAPTIVE CONTROL TECHNIQUE FOR THE SIMULATION OF SEISMIC ACTIONS

This paper describes an interesting test procedure for simulating earthquake-like input to test specimens. The "self tuning" feature of the system raised some questions about how results from reinforced concrete models would compare with performance of the prototype structure. The experimental rig might tend to obscure the effects of cracking that occur in a concrete structure. Consequently, response of the model might be expected to overestimate response of the full-scale structure. The authors reported results tend to indicate this over-estimation does occur.

CONCLUDING REMARKS

In this section several new procedures for making model concrete and small diameter reinforcement are presented. Among the more innovative schemes described are the use of coated aggregates in an attempt to reduce tensile strength of small aggregate concretes and a device that rolls rather than knurls the small diameter reinforcement.

Applicability of microconcrete models to determining shear strength of concrete slabs is reviewed by two authors. The results continue to indicate that care should be taken in trying to extrapolate the results of small scale shear tests to the performance of a prototype. Suitable methods for extrapolating shear strength results have not yet been established. Consequently, large models still are preferred for determination of shear strength of a slab.

Work on modelling steel and steel composite structures is also reported in this section. Of particular interest is the use of vinyls to model steel tanks. White demonstrated the viability of these "elastic" models.

The contribution by Wright, Harding and Evans toward evaluation of shear stud connectors provides useful suggestions. Their results provide two methods for improved calibration of the design procedure currently being considered.

The papers on modelling fibre reinforced microconcrete and on providing discrete confinement for concrete in compression offer interesting possibilities that leave many questions yet to be answered. The apparent difficulty in obtaining similitude in fibres suggests that rather large-scale models still provide the best method for determining performance. The use of bolts to confine concrete appears to work quite well in small specimens but lacks verification at large scale.

Overall, papers in this section provided additional steps toward improved modelling techniques. With stronger reliance on related past work reported in the literature, these results have the potential for providing even more reliable extrapolation of model results to design.

9 Comparison between reinforced concrete prototypes and micro concrete models

J. E. GIBSON
Department of Civil Engineering, City University, London, UK

SUMMARY

In this paper the accuracy in predicting the failure mechanism and the collapse load in the prototype reinforced concrete structure (as constructed the laboratory) from the experimental results obtained from testing a micro concrete model is investigated.

To this end a reinforced concrete prototype structure has been chosen in which an investigation of a certain aspect of slab column behaviour was examined and then the structure was loaded to collapse.

In the prototype structure, a standard concrete mix of normal sand, cement and aggregate with commercially standard reinforcement was used in the casting.

In the construction of the model of the above prototype a standard micro concrete satisfying the physical properties required was used.

The overall scale of the model was chosen so that the main reinforcement diameter was never less than that of the minimum commercially produced size wherever possible.

Deflections and collapse loads in the model were scaled up using model laws predict the deflection and collapse load in the prototype. The agreement between these predictions from the model values and the actual values obtained from prototype tests gave a measure of the reliability of the model tests in predicting behaviour of the prototype structure.

SLAB COLUMN PROTOTYPE

This slab column prototype was in fact constructed to examine the stress distribution and load carrying capacity of a slab column junction for a particular type of design. A standard concrete was selected giving a crushing strength of 42 N/mm^2 at 28 days. The main reinforcement was of 16 mm diameter and was used both in the slab and the column.

The actual prototype is shown in elevation and plan in Figs. (1) and (2) and in detail in Fig. (3) showing prototype, loading frame, rams and pumps. It proved convenient to actually construct the prototype within the testing frame rather than in the concrete laboratory as this ensured that any damage to the structure during transportation from the concrete laboratory to the structures laboratory was avoided. Further, it allowed the column base and the base that supported the other end of the slab to be conveniently located and bolted into position. Having wired the main reinforcement in the slab, strain gauges were located in selected positions in the vicinity of the column slab junction.

The set up for the column slab prototype within the testing frame is shown in Fig.(1). The initial test was to investigate the stress distribution around the column slab junction and thus column loads were applied by the loading ram R1 through a load cell (4) whilst moments at the junction were attained by the loading ram R2 mounted on a load cell (3) located on the slab proper. The end reaction on the slab was recorded by load cells (1) and (2). The load cells (1),(2) and (3) were mounted on needle bearings so that lateral movement could take place.

The complete testing arrangement is shown in Fig. (3) in which the loading rams and load cells may be seen. Pressure was applied to the rams by standard pumps connected to Budenberg gauges and overall load monitoring was achieved by the load cells. Deflections were recorded by dial gauges graduated in 0.01 mm and shown in Figs. (1) and (3). The positions of these are indicated in square brackets throughout the figures, namely [1] to [4].

Having completed the stress investigation of the column slab junction, the author decided to continue loading the structure until structural collapse occurred and it is the experimental results of this latter test which will be used as prototype values to see if in effect they can be predicted by a model test. To this end the structure was retested with a constant load of 24 tons on the column and a distributed load on the slab which was increased incrementally until structural failure occurred. As each increment of load was applied deflections and load cell readings were recorded. This incremental load was distributed to the slab by placing the ram on a spreader beam which ha a cross sectional area of 609 mm by 114 mm, i.e. an area of 69608 mm^2.

The load deflection curves for the four gauges [1] to [4] are shown in Fig.(6) From these it may be seen that certainly over the loading range considered the prototype structure is behaving in a linear elastic fashion. The variation of deflection along the centre line of the slab for different loads is again show in Fig. (5).

Finally the slab failed rather suddenly at a load of 23.4 tons, the nature of the failure being that of a form of shear as may be seen from Fig. (4) in which the distinctive shear crack normally associated with shear failure in beams may be clearly seen. The deflection curves have already shown that up to a load of 22 t (Fig.(6)) the structure was behaving linearly and that or slight increase of load by an amount of 1.4 tons caused failure.

As previously indicated the size of the structure was sufficiently large to allow normal commercial reinforcement and a standard concrete mix to be used its construction. Again it was sufficiently large to be considered as a prototype structure, and that the scaled results from the experimental values obtained by testing a scaled micro concrete model should be able to predict t structural behaviour of this prototype.

SLAB COLUMN MODEL

Inspection of the prototype structure revealed that it would be relatively ea to construct a model to a scale of about 1:2 and still using this scale be ab to use a commercial standard bar diameter together with a micro concrete mix. Thus a 8 mm dia. bar was chosen and a suitable mix to give 42 N/mm^2 at 28 days was selected. The geometry of the model was identical to that of the prototype in all respects but at half scale.

The layout of the reinforcement of the model was identical to that of the prototype and shuttering was identical preserving in detail the secondary transverse beams present in the prototype. Whereas the prototype had been cast within the large testing frame for convenience, the model was cast in its entirety in the concrete laboratory and being of half scale and thus of necessity only one eighth of the weight was easily transportable to the Structures Laboratory for testing.

The model was then loaded in a manner similar to that carried out on the prototype, load cell readings and deflections being recorded as before.

In order to utilise the standard model scales it was necessary to maintain the same state of stress in the model as existed in the prototype. Thus in the prototype a load of 24 tons had been applied axially to the column and thus the corresponding load on the model had to be $24/n^2$ i.e. $24/4 = 6$ tons. This was applied to the column by a 10 ton ram operated by a hand pump and monitored by a Budenberg pressure gauge. This load had of course very little influence on the slab proper but was introduced to stabilise the column and to maintain the exact loading system that operated on the prototype.

The load on the model slab was applied through a spreader beam of dimensions 305 mm by 57 mm, i.e. some 17385 mm^2 and this was loaded incrementally by a second 10 ton ram operated by a hand pump, the pressure being accurately recorded on an extremely accurate Budenberg gauge. The loading pattern applied to the model was similar to that on the prototype and loads and deflections on the four dial gauges were recorded.

The deflection profile across the model centre line is shown in Fig. (7) and the variation in deflection for increasing load of all four gauges is shown in Fig. (8) where again for the range considered the load deflection relationship was obviously linear. Failure again was sudden at a load of 5.5 tons.

Thus, sufficient evidence from the model was now available to predict the deflections and collapse load in the prototype and to then compare these with the experimental values obtained for the prototype. To this end the following analysis was carried out.

MODEL LAW INVESTIGATION

In the following analysis the suffix p appertains to the prototype structure and m to the model structure. The basis of the model investigation was that geometric similarity was preserved throughout and that n was the model scale, i.e. all dimensions obeyed the following law:

(1) $L_p = nL_m$ in which L_p is prototype dimension
L_m is model dimension

That as the structure was of the slab or plate type the deflection δ would obey a law

(2) $\delta \alpha\ q\ a^2b^2/Et^3$ or $\delta = K\ q\ a^2b^2/Et^3$

in which a and b were dimensions of plate
q was applied load/unit area on surface
t was thickness of plate
E is Young's Modulus
K is a contant.

Thus for the model

(3) $\delta_m = K\ q_m\ a_m^2 b_m^2 / E_m t_m^3$

and for the prototype

(4) $\delta_p = K\ q_p\ a_p^2 b_p^2 / E_p t_p^3$

whence from (3) and (4) observing that

$a_p = na_m, \quad b_p = n\ b_m, \quad t_p = n\ t_m$

$$\delta_p/\delta_m = \frac{q_p}{q_m}\left(\frac{a_p}{a_m}\right)^2 \left(\frac{b_p}{b_m}\right)^2 \cdot \frac{E_m}{E_p} \cdot \left(\frac{t_m}{t_p}\right)^3$$

(5) $\dfrac{\delta_p}{\delta_n} = \dfrac{q_p}{q_m} \cdot n \cdot \dfrac{E_m}{E_p}$

Now on the basis that with equal strength mixes at 28 days the moduli would be approximately equal, ie. $E_m \simeq E_p$, then the above low reduces to for

(6) $\quad E_m = E_p$

(7) $\quad \dfrac{\delta_p}{\delta_m} = n \dfrac{q_p}{q_m}$

Now it will be observed that the spreader beam dimensions on the model and prototype were scaled, i.e. $L_p = n\, L_m$ and thus the areas of the spreader beam namely $A_p \alpha L_p^2$ and $A_m \alpha L_m^2$

(8) $\quad A_p = K\, L_p^2 \qquad A_m = K\, L_m^2$

Again if q_p and q_m are the applied surface loads on the slabs through the spreader beams carrying loads W_p and W_m then

(9) $\quad \begin{aligned} q_p &= W_p/A_p = W_p/K\, L_p^2 \\ q_m &= W_m/A_m = W_m/K\, L_m^2 \end{aligned}$

(10) where $\quad \dfrac{q_p}{q_m} = \dfrac{W_p}{W_m} \dfrac{L_m}{L_p}^2 = \dfrac{1}{n^2} \dfrac{W_p}{W_m}$

(11) $\quad \dfrac{\delta_p}{\delta_m} = n \cdot \dfrac{q_p}{q_m} = \dfrac{1}{n} \dfrac{W_p}{W_m}$

(12) and therefore $\delta_p = \dfrac{1}{n} \dfrac{W_p}{W_m} \cdot \delta_m$

Similarly it is easy to show that the stresses in the prototype f_p and the stresses in the model f_m are related by

(13) $\quad \dfrac{f_p}{f_m} = \dfrac{1}{n^2} \dfrac{W_p}{W_m}$

Thus if the model and prototype are of the same material having thus the same physical properties and the failure stress for the material is f_F the in the case of the model the failure stress will occur when $f_m = f_F$ at a model load of

W'_m. In the case of the prototype again failure will occur at a stress when $f_p = f_F$ and the predicted failure load W'_p for the prototype will be given by

$$\frac{f_m}{f_p} = \frac{1}{n^2} \frac{W'_p}{W'_m} \quad \text{with } f_m = f_F \text{ and } f_p = f_F$$

i.e. $\dfrac{f_m}{f_p} = \dfrac{f_F}{f_F} = 1 = \dfrac{1}{n^2} \dfrac{W'_p}{W'_m}$

(14) i.e. $W'_p = n^2 W'_m$

It is interesting to note that the same sort of model laws made be used in examining thin shell structures using similar assumptions.

APPLICATION TO PROTOTYPE

One can now examine whether it is possible to predict the deflections and load carrying capacity of a prototype structure by applying model laws to experimental values obtained from tests on geometrically similar models and a chosen scaling factor n.

The model deflections δ_m given in Fig.(9)(b) were plotted from the following experimental data in column (b) in Table 1, for a model load W_m = 5 tons.

a	b	c	d
Gauge Position	δ_m	$\delta_p = 2.2 \cdot \delta_m$	Experimental δ_p
1	136	299	320
2	180	396	432
3	169	372	451
4	163	359	349

Table 1

For a prototype load of W_p = 22 tons, the predicted deflection values for the prototype δ_p would be given by eqn.(10) namely

$$\delta_p = \frac{1}{n} \cdot \frac{W_p}{W_m} \cdot \delta_m$$

But n = 2 and W_m = 5 tons and W_p = 22 tons

$\delta_p = 2.2\ \delta_m$

Whence, multiplying column (b) by 2.2 we derive the predicted deflections for the prototypes in column (c) in Table 1 and these are shown as the dashed line in Fig. (9)(a). The actual experimental deflections at a load of 22 tons are given in column (d) in Table 1 and as the full line in Fig. (9)(a). Finally the predicted failure load of the prototype can be obtained from equation (14), namely

$W'_p = n^2 W'_m = 2^2 . 5.5 =$ 22 tons

as against an actual experimental failure load of the prototype of 23.4 tons.

Thus in this case the use of linear elastic model laws seemed not unreasonable it being recalled that in this case of shear failure the model remained linear elastic right up to failure. Further experiments involving model and full scale column tests are being persued together with a comparison of a continuous four cell box bridge model with a prototype investigation carried out in the States. It is hoped that these latter investigations will produce further clarification or model/prototype relationships in micro concrete.

REFERENCES

1. Gibson, J.E. "The Design of Shell Roofs", E. & F.N. Spon, Ltd. London, 1968, Ch. 14, Model Investigations.

2. Gibson, J.E. "Computing in Structural Engineering", Applied Science Publishers, London 1975, Ch. 13 Shell and Plate Investigations.

3. Gibson, J.E. "Thin Shells", Pergamon Press, Oxford 1980, Ch. 10 Experimental Investigation of Shells.

Fig. (1) Elevation of Prototype Slab Column

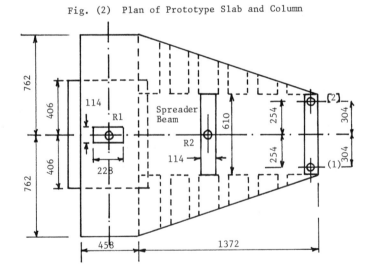

Fig. (2) Plan of Prototype Slab and Column

Fig. (3) Prototype and Loading System

Fig. (4) Failure Cracks in Prototype

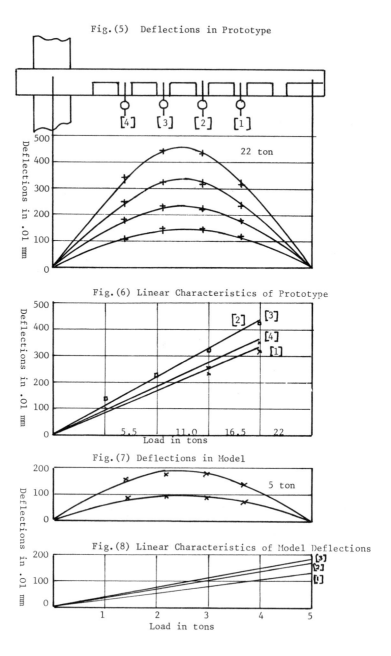

Fig.(5) Deflections in Prototype

Fig.(6) Linear Characteristics of Prototype

Fig.(7) Deflections in Model

Fig.(8) Linear Characteristics of Model Deflections

Fig. (9)(a) Prototype Deflections and Predicted Deflections

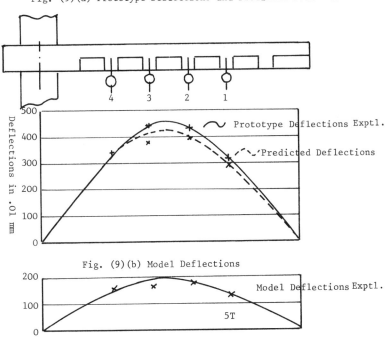

Fig. (9)(b) Model Deflections

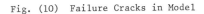

Fig. (10) Failure Cracks in Model

DISCUSSION

Dr F K Garas   Taylor Woodrow Construction Ltd   UK

What information was gained from the micro concrete model test for the National Westminster Tower that could not be obtained from a finite element analysis?

Author's Reply.

It must be recalled that the initial perspex model of the central main core of the National Westminster Tower was used to analyse the main stress distribution in the core walls for the linear elastic range and especially around the openings in the core which formed the entrances from the central lift system to the surrounding office decks. Even in this case it was evident that to gain the same information from a finite element analysis would have involved using a prohibitive number of elements.

When the design changed from the steel truss system at three levels which were to support the office floors, to the base reinforced concrete cantilever lobe system to support all floors, it was considered necessary to determine the ultimate load carrying capacity of these lobes. It was obvious that a perspex model would be useless for this type of investigation, its structural behaviour in the non-linear range being entirely different to that of reinforced concrete. Furthermore, it was decreed that the reinforcement in the lobes should be accurately modelled. This latter point involved the modelling of extremely intricate reinforcement in the deep beams forming the lobes and indeed the provision of mechanical anchorage points for this reinforcement in the core wall itself. The density of the reinforcement was also extremely high.

Given these circumstances the Consulting Engineers required to know:

1) The stresses in the steel reinforcement in the deep shear beams in the cantilever lobes and at other selected points.

2) The load deflection characteristics at various selected points in the lobes.

3) The structural behaviour up to collapse and the ultimate load at structural failure.

These were successfully achieved during the testing of the model.

Given the extreme complexity of the reinforcement and the nature of its anchorage to achieve sufficient bond, and the geometrical complexity of the model, it is extremely doubtful if a finite element analysis could have produced the required information - certainly at the time of the test - some 12 years ago.

10 An evaluation of the factors contributing to size effects in concrete

P. CHANA
Cement and Concrete Association, UK

SUMMARY

This paper is an analysis of the important factors contributing to size effects in concrete. The effects of experimental inaccuracies and strain gradient are discussed. It is suggested that the size effect is greatly influenced by the development and propagation of cracks in the concrete. The relevance of established statistical and fracture mechanics theories is briefly discussed.

INTRODUCTION

Models have been used in the last thirty years in research on structural concrete and for development work in complex structures which cannot be rationally analysed by theory. In recent years, the role of physical models has undergone a decisive metamorphosis. Models are now widely used as a design tool for yielding information on a whole variety of structural responses such as deflections, ultimate loads, cracking and modes of failure. This trend has been accentuated by the adoption of limit state methods of design, which are now also concerned with serviceability conditions such as cracking and deflections, as well as ultimate. Strength models obviously have great potential in this field, provided they can adequately simulate both the elastic and inelastic response of the structure.

Clearly, the model testing engineer must, at some stage, convert his model test results into values reflecting the behaviour of the prototype. Unfortunately, a proper scientific basis has not yet been established for

modelling all aspect   in reinforced concrete. The
situation is further   ffects' i.e.
change in indicated   size. The size
effect may not be im   r research
purposes, but will b   del test results
have to be extrapolated to assess the prototype strength. This can be done
either with the aid of empirical correction factors deduced from repeated
experiments at different scales, or, where no such data are available, by
exercising engineering judgement. A review of experimental studies on size
effects can be found in various references (1,2) and is not repeated here.
This paper is concerned primarily with an analysis of the fundamental reasons
for this effect and an assessment of their significance on model analysis and
design. It is clear that an appreciation of the important factors
contributing to size effects in model concretes will be useful when
extrapolating data from tests on models to represent prototype values.

It is important to be able to distinguish clearly between the possible
reasons for the observed size effects. These can be identified into two
separate groups as follows:

(1) The physical conditions under which the model experiment ought to take
     place, cannot be achieved for technical reasons. This includes
     experimental sources of error in the casting, curing and testing. It
     should be possible to minimise the size effects due to this group by
     using improved experimental techniques.

(2) The laws of similitude used for making the comparison between model and
     prototype are incomplete i.e. one or more significant physical
     parameters cannot be scaled. For example, models have enhanced strain
     gradients compared to a prototype and this will lead to the structural
     properties of the model being altered as will be discussed later in the
     paper. The model testing engineer has little control over these
     factors; it is, therefore, particularly important to be aware of them
     and make adjustments to the test results, as necessary.

EXPERIMENTAL INACCURACIES

Sabnis and Aroni (2) have discussed various experimental sources of error
leading to size effects. Some of these are summarised below.

(a) Fabrication of specimens

As the model scale is reduced, it becomes difficult to maintain the required dimensional tolerances in construction. For very small structures, measurements of the actual dimensions of the completed structure may be necessary in the analysis of results.

The assembly of reinforcement cage may be done by welding, soldering or tying with fine wire. The heating involved in the first two methods may cause local changes in the properties of the reinforcement which could affect the structural response. The use of tying wire may not produce cages of the desired rigidity.

It is well known that the degree of compaction affects the strength. If the compaction procedures followed are the same, i.e. time and frequency of vibration, then smaller specimens will achieve better compaction and consequently show higher strengths. However, if full compaction is achieved in both the models and prototype, the size effect due to this factor can be eliminated.

(b) Curing of specimens

The surface to volume ratio of a specimen varies inversely with the size; hence, curing of models will take place at a faster rate and lead to an earlier gain in strength. However, if wet curing is employed the strength gains should be the same for both models and prototypes.

(c) Testing of specimens

It has been demonstrated that high strain rates lead to higher strengths (3). In a testing machine the rate of cross-head movement should be adjusted according to the size of the specimen. However, for very small specimens, this may not be possible on certain machines, giving artificially high strength values. The use of modern servo-controlled loading machines should lead to a more accurate control of loads giving better results.

STRAIN GRADIENT EFFECTS

The sensitivity of concrete to a strain gradient is well known as a possible reason for the discrepancy between the tensile strength as obtained from a

direct tensile test and a modulus of rupture test (4). The effect of
transverse strain gradient upon the tensile strain capacity has been studied
quantitatively by Blackman et al (5). To isolate the strain gradient effect
from a size effect, tests were carried out on concrete prisms of the same
size but subjected to different combinations of axial, eccentric and flexural
loads, giving different strain gradients. The relation between the ultimate
tensile strain and the strain gradient is shown in Figure 1. It is clear
that the observed increase in modulus of rupture with decreasing depth and
the later onset of cracking in model beams can partly be attributed to a
strain gradient effect. As the depth of the member decreases, the strain
gradient increases, resulting in a higher cracking strain.

In a recent study, the author has suggested that the size effects observed in
the shear behaviour of reinforced concrete can also be attributed to strain
gradient effects (6). Tests were carried out on 400 mm by 200 mm wide
prototype beams, similar half-depth beams and 1:2, 1:3.3 and 1:8.5 scale
model beams. (See Figure 2). The size of the aggregate was also scaled.
The beams had a longitudinal steel percentage of 1.8 with no shear
reinforcement and a shear span to depth ratio of 3.0. The size effect
observed from this test series is shown in Figure 3.

In all the beams tested the majority of shear cracks formed as extensions of
existing flexural cracks. Flexural cracking and the consequent
redistribution of stresses have a profound effect upon the formation of
flexural shear cracks. It follows that delayed flexural cracking will lead
to the formation of a shear crack at a comparatively later load stage.
Table 1 demonstrates that a major proportion of the ultimate shear strength
scale effect is due to 'elastic loading' i.e. from zero load to first
cracking. The 'residual shear strength' or 'post-cracking shear strength',
defined as the difference between the ultimate load and the first cracking
load, is almost constant irrespective of scale.

STATISTICAL THEORIES

The heterogenous nature of concrete has led to attempts being made to explain
size effects in terms of established statistical concepts. The oldest theory
is the 'weakest link theory' of Weibull (7) which assumed that the beam
consists of a number of elements whose strength properties are randomly
distributed. The beam fails when any element is subjected to a stress equal

to its own strength. By using statistical methods of analysis, it is possible to predict the strength of a beam containing a large enough number of these elements. The larger the size, the greater the number of primary elements and the greater is the probability of finding a lower strength element. Thus, the implication is that the strength of a beam will depend on the stressed volume. Modifications of Weibull's theory have been proposed by Tucker (8) and Nielsen (9).

These statistical theories imply failure of concrete in a brittle manner at a point or plane in the specimen. A considerable amount of test work has been done on the actual mechanical behaviour of concrete under tensile stress using very stiff testing machines (10,11). This work has shown that the load-extension curve for concrete under tension has a distinct falling branch and indicates an effectively plastic behaviour. Since considerable deformation takes place at the failure plane before maximum load is reached, the stress over the whole cross-section can be redistributed. The formation of microcracks in the specimen does not necessarily lead to fracture: hence, the 'weakest link theory' or the other related statistical theories cannot be used to explain the behaviour of concrete. It is clear that further work into the failure mechanisms of concrete could lead to a better understanding of size effects. One aspect of work done in this field is discussed briefly in the next section.

FRACTURE MECHANICS OF CONCRETE

Theories related to the fracture mechanics of concrete have recently been used successfully to accurately predict size effects. In particular, a 'fictitious crack model' for concrete has been developed at the University of Lund, Sweden (12, 13). This model considers a zone of failure where plastic deformation takes place during crack growth. The mechanical behaviour in tension is described by means of two curves:

(1) A stress-strain ($\sigma - \varepsilon$) curve, which consists of an ascending branch and possible unloading branches. The stress-strain curve normally does not contain any part with $d\sigma/d\varepsilon < 0$.

(2) A stress-deformation ($\sigma - w$) curve, which consists of a descending branch (and possible unloading branches).

This model can be applied to predict the shear strength of reinforced concrete using finite element techniques. Figure 4 shows the author's test results (6) with the predicted values (14). It is clear that, although the theoretical shear strength curve is lower than the experimental curve, the magnitude of the size effect has been predicted accurately. The lower values obtained for the theoretical shear strength are possibly due to dowel resistance being ignored. This model has also given good agreement with some of the strain gradient effects discussed previously.

DISCUSSION

It is clear that size effects are closely related to the fracture behaviour of the model material. In the analysis of model test data, it is not appropriate to apply a blanket factor of safety of, say 1.5, on the member resistance. The mode of failure is an important factor to be considered. Failure modes associated with tensile material fracture are particularly susceptible to size effects and care should be taken in such cases in assessing model test data. It is hoped that this analysis of size effects will act as a stimulus in the development of new model materials giving better correlation with prototype behaviour.

At first sight, it would appear that the use of model analysis should be discouraged for solving problems due to uncertainties of size effects. Clearly, the model testing engineer must exercise judgement in selecting the parts of the structure that should be simulated as closely as possible and those where 'distortion' will have minimum effect on the measurements being made. The model scale has to be chosen carefully and the results adjusted accordingly. However, this situation is not so different from that of a structural engineer who employs a mathematical model to design a structure. The engineer makes basic assumptions with respect to boundary conditions, material properties and structural interaction. Allowance has then to be made for assumptions made during analysis, variations in material strength and construction tolerances by applying safety factors. Thus, the uncertainties of design values obtained by model testing or structural analysis would appear to be similar; the fields of application of one or the other method depend largely on the circumstances of the particular design problem. Models will continue to be a useful tool in solving complex design problems, provided due attention is paid to the similitude requirements and any size effects in the analysis of results.

Table 1 Comparison of shear stress at first cracking and ultimate load

Beam scale	A Ultimate shear stress (N/mm^2)	Load at onset of cracking as a percentage of ultimate load	B Shear stress at onset of cracking (N/mm^2)	'Residual' shear stress = A - B (N/mm^2)
1:1	1.08	44	0.48	0.60
1:2	1.35	49	0.66	0.69
1:3.3	1.56	58.5	0.91	0.65
1:8.5	2.04	57	1.37	0.67

REFERENCES

1. SABNIS, G M, and MIRZA, S M, 'Size effects in model concretes?' Proceedings of the American Society of Civil Engineers, Vol. 105, No ST 6, (June 1979).

2. SABNIS, G M, and ARONI, S 'Size effects in material systems - The State of the Art' International Conference on structure, solid mechanics and engineering design at University of Southampton, England, (April 1969).

3. WRIGHT, P J F, 'The effect of method of test on the flexural strength of concrete', Magazine of concrete research, No 11, (October 1952).

4. PRICE, W H, 'Factors influencing concrete strength', Journal of the American Concrete Institute, Proceedings Vol 47, No 6, pp 417-432, (February 1951).

5. BLACKMAN, J S, SMITH, G M, and YOUNG, LEE, 'Stress distribution affects ultimate tensile strength', Journal of the American Concrete Institute, Proceedings Vol 55, No 6, (December 1958).

6. CHANA, P S, 'Some aspects of modelling the behaviour of reinforced concrete under shear loading', Cement and Concrete Association Technical Report 543, Wexham Springs, UK, (July 1981).

7. WEIBULL, W, 'A statistical theory of the strength of materials', Proceedings of the Royal Swedish Institute of Engineering Research (1939).

8. TUCKER, J, 'Statistical theory of the effect of dimensions and of method of loading upon the modulus of rupture of beams,' Philadelphia, Proceedings of the American Society for Testing Materials, Vol 41, (1941).

9. KELLERMAN, W R, 'Effect of size of specimen, size of aggregate and method of loading upon the uniformity of flexural strength tests', Public Roads, Vol 13, (1933).

10. MARATHE, M, 'Stress-strain characteristics of concrete in tension,' Thesis submitted to the University of Leeds for the degree of Doctor of Philosophy, (August 1967).

11. HERTZBERG, R W, 'Deformation and fracture mechanics of engineering

<u>materials</u>,' John Wiley and Sons, New York (1977).

12. HILLERBORG, A, 'The application of fracture mechanics to concrete', Conference on contemporary European concrete research, Stockholm, (9-11 June 1981).

13. MODEER, M, 'A fracture mechanics approach to failure analysis of concrete material', Report TUBM-1001, University of Lund, Sweden, (1979).

14. HILLERBORG, A, and GUSTAFFSON P J, Private communication, (November 1981).

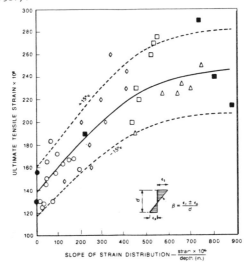

FIGURE 1:- Effect of strain distribution on ultimate tensile strain (Ref. 5)

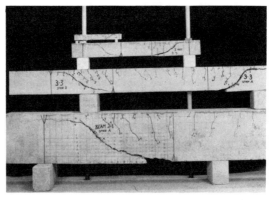

FIGURE 2:- Range of beam sizes tested

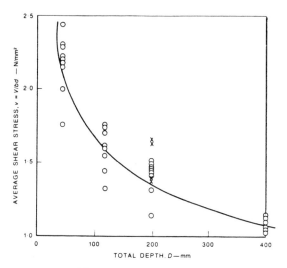

FIGURE 3:- Test results

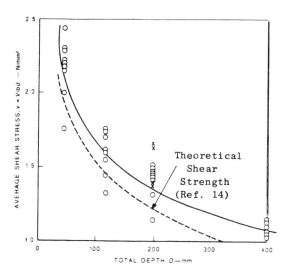

FIGURE 4:- Comparison of test results with the 'fictitious crack model'

DISCUSSION

Dr R J Cope   University of Liverpool   UK

In addition to the size effects outlined by Mr Chana, it should be mentioned that the distribution of concrete strength over the thickness of prototype and model beams can be very different. In general, a thick member has a decrease in strength towards the top. This will affect the contribution of both the aggregate interlock and compression block contributions to the shear strength. It is difficult to see how this strength variation could be simulated in model beams, and how the results could be extrapolated if it is not represented.

In the paper, the "weakest link theory" to explain the behaviour of concrete is dismissed because concrete has been shown to have a falling branch to the tensile stress-strain curve. However, such curves are only obtained using very stiff testing machines. As concrete in a beam is not surrounded by an immensely stiff structure, perhaps the falling curve does not apply to practical structures. The statistical theories would then be relevant.

In Table 1, results are presented that show the residual shear capacity after the onset of cracking is sensibly constant. As the cracked beam has to carry the entire shear force, could Mr Chana please explain whether these numbers have any physical significance? Is the post-cracking load path different in differently-sized members?

As tensile cracking is delayed in model beams, the relative, sudden energy release will be greater than in a deep beam. This could result in relatively greater bond slip and crack widths in model beams which could, in turn, affect the shear capacity. Has the author noticed such effects in his tests?

Author's Reply

Dr Cope has raised some interesting points which could not be properly covered in the paper owing to constraints on space. The points made by him are discussed, in turn, below.

Dr Cope is right in suggesting that there is a strength gradient effect in deep beams. For the type of flexural shear failures discussed in the paper, failure cannot be attributed to any distress in the compression zone. In the author's opinion, this effect is not significant for flexural shear failures. It is, of course, an important factor to be considered when considering other aspects of concrete behaviour such as shear compression failures, compressive failure of over-reinforced beams and bond stress of top cast bars.

It is true that a falling branch in the tensile stress-strain curve is only obtained using very stiff testing machines. This example was given in the paper to illustrate the ductility of concrete in tension. The cracking behaviour in practical structures also exhibits considerable ductility.

Statistical theories imply failure of concrete in a brittle manner at a point or plane in the specimen. In concrete, most cracks are, in fact, stable. The formation of a micro-crack does not do much more than to reduce a local stress-concentration. The result is that many cracks are initiated in the concrete before the maximum fracture stress has been applied. Any theory of failure should consider a finite zone of failure taking progressive collapse into account.

For flexural shear failures, flexural cracking and the consequent redistribution of stress has a profound effect upon the formation of shear cracks (see, for example, Hamadi Y D and Regan P E 'Behaviour in shear of beams with flexural cracks', Magazine of concrete research, Vol 32, No 111, June 1980). Delayed flexural cracking will lead to the formation of a shear crack at a comparatively later load stage. This is somewhat analogous to prestressed concrete design where flexural cracks are delayed due to the effect of the prestress leading to an enhanced shear capacity.

Crack widths were not directly measured during the tests owing to practical difficulties in obtaining meaningful results. However, the table below gives values for average spacing and number of cracks in the shear span for all the beams tested. It is clear that fewer cracks are formed in the small-scale beams implying a relative increase in crack width. As pointed out by Dr Cope, this could be due to a relatively greater energy release to cause cracking.

Cracking similitude (average values)

Beam scale	Average spacing (mm) (scaled spacing in parenthesis)	Number of cracks in shear span	Load at onset of cracking as a percentage of ultimate load
1:1	115	7	44
1:2	60 (57.5)	6	49
1:3.3	41 (34)	5	58.5
1:8.5	14.4 (13.5)	5	67

Mr M J G Connell   PSA   LONDON   UK

Table 1 and Figs 3 and 4 imply that for an approximate scale difference of say 10:1, the ultimate shear stress varies by 100%+. It would appear that for any 1/10 scale test that is liable to fail through shear, the results will be +100% too high. Is this true? Perhaps the author of Paper No 11 would comment too as his tests did not seem to give such a large difference, although he does say that flexural failure scales satisfactorily and his tests appear to have a size difference of about 4:1 which should be alright for trials.

Author's Reply.

Mr Connell is right in suggesting that the ultimate shear stress for a 1/10 scale specimen (ie 40 mm deep) would be over double the shear resistance of the 400 mm prototype. The scale effect in shear is particularly evident for small sizes. In the author's opinion, it is more helpful to consider this phenomenon in terms of absolute size rather than scale which is a relative quantity. The example given by Mr Connell is for a beam 40 mm deep which would not be considered to be a reasonable size for model shear studies under any circumstances. A model of this size requires special materials and is largely influenced by strain gradient effects. In such a model, the failure mode could well be flexure as opposed to shear failures in the large scale models.

Dr F K Garas   Taylor Woodrow Construction   UK

Why do size affects matter? In your model work, if you try to simulate faithfully everything as near as you can and then relate the experimental results to the theoretical prediction, surely you have achieved your orbjectives? The next step would be to apply the mathematical model to predict the behaviour of the full scale structure.

Author's Reply

The point raised by Dr Garas is an interesting one. Physical models may be used to check the validity of analytical models as is commonly done in research laboratories. In many cases, the analytical model can be used with confidence at full scale. However, the phenomenon of size effects is usually not implicit in the analytical model and the problem of size affects still remains. To illustrate this, reference will be made to the design rules for shear in the United Kingdom bridge code, BS 5400 (1984 edition). The shear resistance of beams without shear reinforcement is considered to be a function of the cube root of both the compressive cube strength and the ratio of longitudinal reinforcement. This is roughly in agreement with various analytical models proposed for the shear resistance and confirmed by tests on laboratory specimens. There is still a need to modify the allowable shear resistance with an empirically obtained size correction factor known as the depth factor, ξ_s, which is given as $\xi_s = \sqrt[4]{500/d}$. This example illustrates that both analytical and physical models have their limitations.

11 Size effects in microconcrete beams and control specimens

S. MAJLESSI, F. A. NOOR and J. B. NEWMAN
North East London Polytechnic, UK

SUMMARY

The results of a detailed investigation into the influence of size on the strength of control specimens in compression and the behaviour of plain and reinforced microconcrete beams in flexure are presented. Control of compaction by the measurement of fresh concrete density has been used. It appears that the difference in compressive strength of control specimens of various size, is mainly due to differences in methods of fabrication, curing and testing. Whilst the flexural test results show a clear size effect for unreinforced beams, this effect is not so obvious in reinforced beams in which the propagation of cracks is arrested by the reinforcement. As the moment-curvature relationship of reinforced specimens of different depth is similar up to yielding of reinforcement, it would appear that the size of model for a given prototype is not critical. However, there is a significant increase in inelastic deformations with strain-gradient and further work is required on this topic as well as the behaviour in shear.

INTRODUCTION

The effect of size on the strength of control specimens and behaviour of microconcrete beams has been the subject of a number of recent research reports[1,2]. Although the volumetric theory[3] may be applicable to some engineering materials, its validity for structural concrete is doubtful. The size and type of aggregate and the methods of compaction, curing and testing may influence the test results. Differential compaction in microconcrete has been suggested[2] as one of the sources of variation in strength. It appears, from the present work, that unless compaction is the same in all specimens and particular care is taken to eliminate the differences in curing and testing, any variation in strength of different sized control specimens cannot really be identified as a true size effect.

Earlier research by one of the authors has shown that the strain gradient across the section of a beam has an influence on flexural tensile strength and inelastic deformations. Unfortunately no attempt was made to isolate the effects of this from that of aggregate type and size, and compaction. In the present study, the aggregate size has been kept constant, the compaction controlled and the variation in strain-gradient obtained by changing the depth of the microconcrete beams from 20 to 80 mm, whilst keeping the width (30 mm) and span/depth ratio (7.5) constant. By observing the complete range of behaviour of model beams it is now possible to specify the depth of model required for a given prototype and the possible size effects which may occur in model tests.

CONTROL SPECIMENS

All specimens were compacted on a constant frequency/variable amplitude vibration table. Initial exploratory tests showed that the mass of mould has an influence on compaction and, to eliminate this effect, the moulds were clamped to the table. Both wet and dry densities were monitored for all specimens and there appeared to be a sufficient correlation between wet density and compressive strength for this to be adopted as the method of compaction control. The next stage of the work involved the determination of the intensity and time of vibration required to obtain full compaction for each size of control specimen. The cube strengths of over 400 specimens available at this stage were subjected to statistical analysis which indicated a probability of over 95% that the strength was repeatable due to control of wet density rather than by pure chance.

Table 1 gives the mix used in the initial study. The curing was carried out under water, at a constant temperature, until just before testing.

Table 1 Details of microconcrete mix

Water-Cement Ratio	Aggregate-Cement Ratio	Aggregate Grading mm		
		1.7-2.3	0.6-1.18	0.15-0.6
0.5	4	40%	20%	40%

* The aggregate used was from Leighton Buzzard Sand.

Figure 1 shows the specially designed lockable spherical seating and the end plates in the cube tests in order to ensure axiality of loading and to minimize the secondary end restraint induced by the testing machine.

Results of Control Specimens

The results of over 400 specimens in series 1 to 4 showed that, apart from a few specimens which had a high vibration input (over compacted), the strengths of cubes with similar wet densities were close for all cube sizes (25, 50 and 75 mm). Typical results for Series 5 are shown in Table 2 from which it can be seen that although the range of cube sizes is much greater, the variation in their strength is still small (±1.5%). There were two other similar groups of tests and all results were in agreement with the theoretical findings of other researchers[4] which indicated that the compressive mechanism of failure in concrete is independent of size, provided all the boundary conditions are controlled. It seems that for determination of concrete strength in models control specimens as small as 25 mm could be used provided special care is taken to ensure that specimens are not damaged at the time of demoulding, since significant imperfections lead to a reduction in strength and an increase in variability. For practical purposes the 50 mm cube is recommended, therefore, as the smallest size for most model work.

Table 2 Series 5 : Microconcrete Cubes, 25 to 150 mm

Group *	Cube Size mm	No. of Specimens	Wet Density kg/m^3	Strength N/mm^2	Coefficient of Variation %
1	25	5	2340	39.48	3.4
	75	3	2340	40.42	0.6
	150	3	2340	39.40	2.6

* Total number of specimens in three groups was 33.

Table 3 gives typical results of control specimens, for groups 1 and 5 of the 8 groups comprising Series 6, made with 20 mm maximum aggregate concrete. Uniform wet density was difficult to achieve in group 1 due to the use of a low workability mix, and the strengths are subject to its variability. This was not a problem in group 5 and, once again, no size effect is apparent. It appears that 75 mm cubes are sufficiently large if the mix has medium to high workability and the maximum aggregate size is not greater than 20 mm. A recent report[5], in which the results of 388 cylinders are presented, makes a similar recommendation.

Table 3   Series 6 : Structural Concrete Cubes, 75 to 150 mm

Group *		Cube Size mm	No. of Specimens	Wet Density kg/m^3	Strength N/mm^2	Coefficient of Variation %
1	Low Workability	75	3	2410	62.70	1.79
		100	3	2390	56.42	1.70
		150	3	2380	55.73	0.71
5	Medium to High Workability	75	3	2360	43.74	1.96
		100	3	2360	43.26	0.81
		150	3	2360	43.47	2.00

* Total number of specimens in 8 groups was 72.

BEAM TESTS

The beams were cast on their sides so that the direction of casting was across the width of the beam (see Figure 2). This technique ensured that uniform vibration was applied across the depth. The moulds were clamped to the vibrating table, and the time and intensity of vibration were controlled such that similar wet densities could be obtained. The weight of fresh concrete compacted into the mould was measured after casting and the volume of specimens measured after hardening using the water displacement technique. The homogenuity of specimens was checked by testing cubes cut from various parts of the beam.

The microconcrete was well graded as shown in Table 1. Model high yield steel, described in another Seminar Paper by Noor et al, was used as the main reinforcement with the percentage used being in proportion to overall depth. Further details are given in Table 4.

Flexural Deformation Measurement

The following three systems were used for measuring and recording the force-deformational behaviour of beams (see Figure 3). System 1 comprised two spring-strain-gauge transducers glued to top and bottom of the beam, by the aid of aluminium pieces. This technique proved to be suitable for collecting data relating to the initial cracking of plain and reinforced beams. The gauges spanned over the constant moment zone, which varied in length from 50 to 200 mm in beams of different depth.

System 2 consisted of two LVDT transducers, connected parallel to, and at a distance from, the beam by the aid of lever arms. This method was used mainly to measure large deformations at yielding and collapse of reinforced beams. System 3 included an LVDT displacement transducer positioned under the beam to record the mid-span deflections throughout the loading range. A Dartec servohydraulic machine was used to apply the load using the transducer of System 3 to give a sensitive feedback displacement control. Signals from both the load and displacement instrumentation were recorded on chart recorders.

Results

The plain beams consisted of 4 sets of small, medium and large specimens and the test results showed that the larger beams failed at a relatively lower flexural strength and deformation. This size effect was not confirmed in the reinforced beams which comprised another 4 sets of three different sized specimens. Half of these were lightly reinforced and the rest heavily reinforced. Whilst the size effects in the moment at cracking and the strength and curvature at yield were small, there appeared to be significant differences in the inelastic deformations. Typical force-deformational relationships for a reinforced beam are given in Figures 4 and 5 and a summary of all results is shown in Table 4. A non-dimensional moment-curvature relationship for the lightly reinforced series is shown in Figure 6. The absence of size effect in strength and curvature at yield is not unexpected since in under-reinforced beams the behaviour up to yielding is influenced more by the properties of the reinforcement than that of concrete.

Table 4 Reinforced Beams

Beam Type		Size	ℓ mm	Bottom Reinforcement diameter mm	f_{cu} N/mm²	f_y N/mm²	Cracking Tensile Strength N/mm²	M_c/M_u	$\dfrac{M_u(\text{exp})}{M_u(\text{Theor})}$	Yielding Curvature $(\times 10^{-6})\phi y.d$	Inelastic Rotation $\div \Delta\delta/\ell$
Lightly Reinforced		Small	50	2 × 1.96	53-55	487	4.98	0.266	1.02	3946	0.107
		Medium	125	5 × 1.96	53-55	487	4.40	0.239	1.01	3544	0.073
		Large	200	8 × 1.96	53-55	487	4.30	0.234	1.01	3512	0.033
Heavily Reinforced		Small	50	2 × 2.85	53-55	574	4.48	0.117	1.11	5543	0.033
		Medium	125	5 × 2.85	53-55	574	4.51	0.130	1.06	5388	0.013
		Large	200	8 × 2.85	53-55	574	4.91	0.138	1.11	5498	0.010

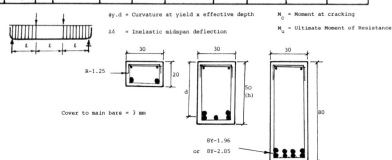

$\phi y.d$ = Curvature at yield × effective depth
$\Delta\delta$ = Inelastic midspan deflection
M_c = Moment at cracking
M_u = Ultimate Moment of Resistance

* Each result is the average for a pair of beams.

Figure 6 shows that the smaller beams have sustained a greater ultimate curvature. This is also demonstrated in Table 4 in which the inelastic rotations of beams reduce with depth. Such a difference may be due to the greater restraint present at the load points in smaller beams and to the transfer of stresses across the narrow cracks immediately below the neutral axis.

CONCLUSIONS

1. For model structures, the size of control specimens need not be as small as the minimum member dimension provided proper care is taken in compaction, curing and testing. For prototype mixes, with medium to high workability, this size can be as small as 75mm which confirms the results of another recent investigation[5].

2. The fresh concrete density method of compaction control should prove useful in the production of similarly compacted laboratory specimens, standardization of the methods of casting cubes and optimizing the compaction of precast concrete units.

3. The flexural strength of plain concrete beams is size dependent and decreases with depth. The influence of size on strength and deformation of model reinforced concrete beams designed to fail in bending is insignificant at the cracking and yielding stages but the inelastic deformations reduce with depth. In view of this it is recommended that further research is carried out to study the effects of size on inelastic deformations and shear behaviour.

REFERENCES

1. WALDRON, P and PERRY, S H, 'Small scale microconcrete Control Specimens', Reinforced and Prestressed Microconcrete Models, Construction Press, (1980).

2. EVANS, D J and CLARKE, J L, 'A comparison between the flexural behaviour of small scale microconcrete beams and that of prototype beams', Cement and Concrete Association, Technical Report 542, (March 1982).

3. WEIBULL, W, 'A statistical theory for strength of materials', Proceedings of the Royal Swedish Institute for Engineering Research No.149, 151 and 153 (1939).

4. JAYATILAKA, A de S and TRUSTRUM, K, 'Mechanism of compression failure in brittle materials', Journal of Material Science, Vol.13 (1978).

5. NASSAR, K W and KENYON, J C, 'Why not 3 x 6 inch cylinders for testing concrete compressive strength', Journal of American Concrete Institute, No.817 (February 1984).

ACKNOWLEDGEMENT

The authors wish to record their gratitude to the Science and Engineering Research Council, which funded part of the work carried out during the past two years.

Figure 1 Lockable spherical seatings for control specimens

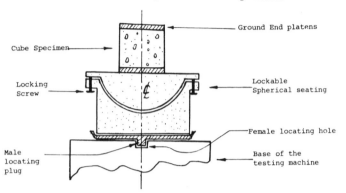

Figure 2 Casting of beams

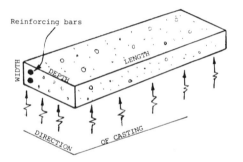

Figure 3 Measurement system for flexural deformations

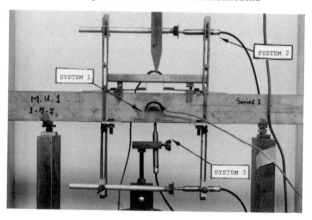

Figure 4 Total load - sum of transducer movements

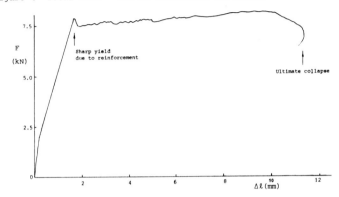

Figure 5 Total load - bottom strain

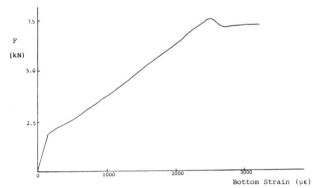

Figure 6 Moment-curvature relationships of lightly reinforced beams

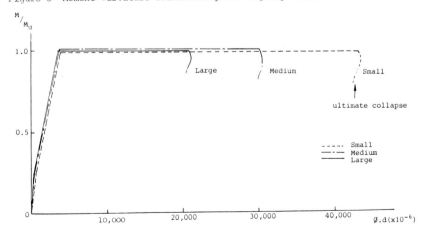

DISCUSSION

Prof Richard N White   Cornell University   USA

This paper presents some very interesting results on size effects in reinforced beams and in control specimens. The use of constant width beams with variable depth is an escellent way to get at the potential problems of size effects.

Studies on model concrete beams made from gypsum concrete at Cornell University (Ref 1, pp 147-150) show that differential density and differential moisture content were the critical factors in size effects. An extensive test series on cylinders ranging in size from 3 by 6 inches to 1 by 2 inches proved that size effects in uniaxial compression and in split cylinder strength disappeared when compaction procedures were used that produced uniform density in the four different size cylinders, and when drying conditions were controlled to provide the same moisture content in all sizes of cylinders. A modest, but definite, size effect still existed in modulus of rupture specimens, however, which leads one to the conclusion that strain gradient is a contributor to observed flexural strength size effect. Achieving uniform density among different size cylinders is a difficult task and must be done by trial and error; there is no way to directly scale the compaction process for either hand compaction or vibratory compaction.

It is suggested that similar studies on Portland cement based model concretes, with a focus on density and moisture content, might provide fruitful results.

The point made by the authors about greater curvature being achieved in the smaller models does not agree with other available evidence. To cite one example, the 1/10 scale models of reinforced concrete beam-column joints presented in Ref 1 (pp 444-460) exhibited moment-rotation responses that were strikingly similar to those of the prototype joints. In this modelling study, great pains were taken to ensure that the model concrete had tensile strength similar to that of the prototype, and the model reinforcing was made of deformed bars that properly simulated the bond behaviour of prototype reinforcement. Actual prototype moment-rotation results are compared with results scaled from the 1/10 scale model in Fig 2 below (taken from Refs 1 and 2). There were no apparent size effects between the prototype and the 1/10 scale model for a severe loading regime that included numerous cycles of fully reversing loads with ductility factors up to 5.

While shallower model beams (that would correspond to the shallowest beams tested by the authors of Paper 11) were not used in this research study, other unpublished results from smaller model beams did not indicate the large changes noted by the authors.

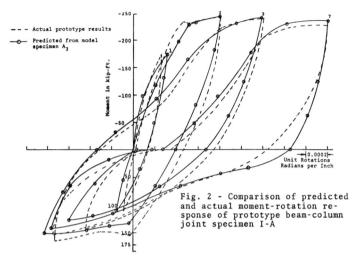

Fig. 2 - Comparison of predicted and actual moment-rotation response of prototype beam-column joint specimen I-A

REFERENCES

(1) SABNIS, GM, HARRIS, HG, WHITE, RN, AND MIRZA, SM, Structural Modeling and Experimental Techniques, Prentice-Hall, 1982.

(2) WHITE, RN AND CHOWDHURY, AH, 'Behaviour of Multi-Storey Reinforced Concrete Frames Subjected to Severe Reversing Loads', Proceedings of the Symposium Resistance and Ultimate Deformability of Structures Acted on By Well Defined Repeated Loads, IABSE, Lisboa, 1973, pp 205-211.

Mr M J G Connell   PSA   LONDON UK

What should the minimum size or depth of concrete beam or slabs be if we are to avoid large shear differences in small scale tests of, say, 1/10 scale or less?

Author's Reply

The model beams tested by us were designed to fail in bending and we concluded that further research is required into the effect of size on the inelastic deformations and shear behaviour. However, the relationship in Fig 3 of Paper 10 is clearly not linear and, therefore, not all 1/10 scale models will have a large size effect. Recent tests at North East London Polytechnic show that the inelastic deformations increase rapidly when the model depth is reduced below 60 mm and it could also be argued that a similar trend is apparent in Fig 3, with respect to shear strength.

It should also be noted that micro-concrete mixes may be designed to give compressive to tensile strength ratios comparable to those of prototype mixes, in which case the size effects are significantly reduced. Please see Ref 5 of Paper 7.

Dr P D Moncarz   Failure Analysis Associates   California USA

There is an apparent inconsistency between the size effect with respect to tensile strength of plain concrete and the calculated tensile stress at cracking shown in Table 4.

Author's Reply.

It would appear that the cracks in reinforced concrete beams are arrested by the steel bars and do not propogate as easily as those in either plain concrete beams or other plain concrete specimens subject to indirect tension.

Dr M EL-Adawy Nassef   Cairo University, Egypt

The grade of F_{cu} = 53-55 N/mm^2 cannot simulate the case of concrete used in most of the ordinary concrete structures. In addition, using such a mix of relatively high modulus of stiffness "E" besides, the size effects of aggregates and specimens considered may make the correlation, between test results and actual elements questionable.

12 Comparisons of tests of medium-scale wall assemblies and a full-scale building

B. J. MORGAN
Construction Technology Laboratories, Illinois, USA
H. HIRAISHI
Building Research Institute, Japan
W. G. CORLEY
Construction Technology Laboratories, Illinois, USA

SUMMARY

The U.S.-Japan Cooperative Research Program on Reinforced Concrete Structures consists of full-scale tests, reduced-scale tests, component tests, and analytical investigations. This paper presents a review of tests of two approximately 1/3-scale planar specimens. Details of construction, load history and test results are presented. Comparison of test results with associated analytical predictions and full-scale test results are given.

HIGHLIGHTS

A planar wall-frame assembly and an isolated wall were constructed and tested under reversing static loads. The wall-frame assembly was a medium-scale representation of the wall-frame section of a full-scale structure tested in Japan. The isolated wall was constructed and tested to simulate the wall section of the wall-frame assembly. Analytically predicted strengths were ten and four percent less than the measured strengths of the wall-frame assembly and the isolated wall, respectively. Behavior of the wall of the wall-frame assembly was adequately predicted from behavior of the isolated wall.

Overall behavior of the medium-scale specimens and the full-scale structure were similar. The full-scale structure was stiffer and stronger than was predicted from scaling up the medium-scale results assuming planar behavior in the direction of load. Increased strength was due to three-dimensional effects and participation of floor slabs (Ref. 1). An analysis made assuming strength contributions of three-dimensional effects gave a predicted full-scale strength that agreed well with measured strength of the full-scale structure.

OBJECTIVE

The work described in this paper is based on "Recommendation for U.S.-Japan Cooperative Research Program Utilizing Large-Scale Testing Facilities (Ref. 2) proposed by U.S.-Japan Planning Group, Cooperative Research Program Utilizing Large-Scale Testing Facilities.

The overall objective of the Planning Group recommendation was "to improve seismic safety practices through studies to determine the relationship among full-scale tests, small-scale tests, component tests, and analytical studies." In keeping with this overall objective, two medium-scale reinforced concrete planar structures were constructed and tested at the Construction Technology Laboratories. A wall-frame assembly, as shown in Fig. 1, and an isolated wall representing the wall section of the assembly were tested (Ref. 1).

Figure 1. Wall Frame Assembly

The specific objectives of the tests of the planar structures were (a) to correlate test results of the two specimens, (b) to correlate test results with analytically predicted behavior to verify analytical and structural modeling techniques, and (c) to provide data for the overall study to determine the relationships from the various testing and analytical approaches.

TEST SPECIMENS

Full-Scale Structure

The full-scale structure was seven stories, 71-1/2 ft (21.8 m), high, three bays long in the direction of load application, and two bays wide. It contained a single structural wall in the center. The specified strength of the concrete used was 3840 psi (26.5 MPa). The specified yield strength of the reinforcement was 49,800 psi (343 MPa).

Wall-Frame

The medium-scale wall-frame specimen consisted of a single planar frame with a structural wall as shown in Fig. 1 (Ref. 1). The ratio of the concrete dimensions of the wall-frame test specimen to the full-scale structure was one to 3.5. This scaling provided a structure approximately 20-1/2 ft (6.2m) high and 16 ft (4.9m) wide from column to column.

Concrete design strength was 4000 psi (27.50 MPa) at 28 days, closely matching the concrete design strength of the full-scale structure. Primary reinforcement in the beams and columns was No. 3 bars conforming to ASTM Designation: A615 Grade 60 (Ref. 3). The percentage reinforcement in the wall frame specimen was chosen to give a stiffness for the planar specimen equivalent to the calculated two-dimensional stiffness of the entire full-scale structure.

Isolated Wall

The isolated wall test specimen duplicated the wall portion of the wall-frame structure except that it did not have slab stubs. The isolated wall also was shorter than the wall-frame specimen. It was loaded to represent the entire seven level wall-frame specimen. The same concrete mix design used for the wall-frame specimen was used for the isolated wall.

TEST PROCEDURE

Full-Scale Structure

Initially, the full-scale structure tested in Japan underwent a series of low level vibration and static tests. These were followed by four tests utilizing a modified pseudo-dynamic test procedure. During this series of tests, the structure was subjected to increasing lateral load cycles until extensive cracking and yielding occurred. After this series, the structure was repaired and tested again. It was then modified with non-structural elements, tested again, and finally tested to maximum strength levels. The medium-scale tests conducted at Construction Technology Laboratories corresponded to the full-scale tests utilizing the modified pseudo-dynamic test procedure.

The modified pseudo-dynamic test approach used the load apparatus at all seven levels. Also, real earthquake time-history data provided basic lateral load input. However, modifications were made to the input function to insure that the structure responded essentially in only its first mode. These modifications retained the earthquake-like load time-history at the roof level, while subjecting the lower levels to an inverted triangular load distribution.

Wall-Frame

The wall-frame specimen was loaded laterally with an inverted triangular distribution. This distribution closely simulated the earthquake-like lateral load used in the test of the full-scale structure. The specimen was loaded alternately in each lateral direction. Vertical load, in addition to the specimens self weight, was required to simulate the axial stresses present in the lower stories of the full-scale structure. Details of loading are described in Ref. 1.

Isolated Wall

The isolated wall was loaded laterally with a single force applied at the top, or fifth story level, of the specimen. This was the same level as the centroid of the inverted triangular load system used for the wall-frame specimen. Consequently, externally applied bending moment and shear at the first story were similar for both specimens (Ref. 1).

TEST RESULTS

Medium-Scale Specimens

Wall-Frame

The mechanism that developed within the wall-frame specimen consisted of hinging in the columns and wall at the base of the structure with all other hinging taking place in the beams. The mechanism was fully developed at a drift ratio of 1.5% measured at the top of the specimen. Maximum lateral load was maintained at this drift. Measured load versus drift is shown in Fig. 2.

Figure 2. Load versus Drift - Wall-Frame

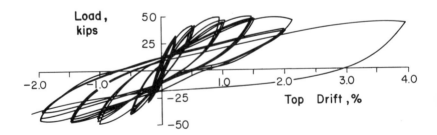

Isolated Wall

The isolated wall reached its maximum lateral load capacity at a drift of approximately 1%. However, significant additional drift was obtained while maintaining lateral load resistance. Measured load versus drift for the isolated wall is shown in Fig. 3.

Figure 3. Load versus Drift - Isolated Wall

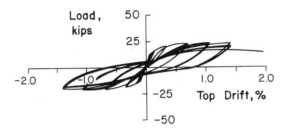

Basic Behavior

During both tests, the wall boundary element that was in tension elongated considerably when compared to the shortening of the boundary element in compression. Moreover, most of this vertical elongation was concentrated in the first story. After yield was well developed, the wall rotated essentially as a rigid body about a pivot point located at the base of the boundary element in compression. The boundary element in tension elongated more or less uniformly from the base to the top of the first story. The developed wall mechanisms were clearly flexural.

Analysis

A structural analysis was made of both specimens utilizing the measured mechanical properties of the materials. Maximum moment capacities of the wall, beams, and columns were determined from a moment versus curvature analysis assuming plane sections remain plane during bending. Strain hardening of the reinforcement was considered. Imposed axial load effects were included in the analysis (Ref. 1). Maximum load capacity of the wall-frame as determined by the test was 1.1 times the analytically predicted maximum load. Maximum test load for the isolated-wall was 1.04 times the analysis value.

Full-Scale Structure

Strength Comparison

The full-scale structure sustained a lateral load of approximately 961 kips at a drift of 1.5%. Measured load versus drift for the full-scale structure is shown in Fig. 4.

A predicted lateral load of 619 kips is obtained by scaling up the results of medium-scale tests to full-scale considering only planar

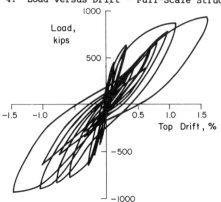

Figure 4. Load versus Drift - Full-Scale Structure

structural behavior. This predicted load is considerably less than the load sustained by the full-scale structure. There is strong evidence to suggest that the increased strength obtained during the full-scale test is due in large measure to three-dimensional effects. The factors that appear to be the major contributors are the transverse beams, the floor slabs, and axial load induced in the wall.

As discussed previously, the boundary element of the structural wall in tension elongated considerably when compared to the shortening of the compressive boundary element. Most of this elongation was concentrated in the first story. As a consequence, beams running transverse to the tension boundary element in the full-scale structure experienced relative upward movement of their ends. This came about because they were connected on one end to the boundary element which was displacing vertically, and on the other end to a conventional frame which experienced little vertical deformation. Transverse beams framed into both sides of the boundary elements of the full-scale structure and therefore contributed to its overall strength.

In the planar wall-frame structure, the slab stubs were comparatively narrow and deformed in one plane only as they contributed to bending strength of the beams. In the full-scale structure, the slab in the vicinity of the tension boundary element deformed in two planes. The slab worked with beams both in the plane of, and transverse to, the structural wall. Therefore, the full-scale structure with slabs deforming in two planes had increased strength over that predicted from the test of a wall-frame specimen with slab stubs deforming in only one plane.

Scaling up the results of the medium-scale tests including the effects of transverse beams, and slabs assumed to be fully effective in negative moment with all hinging beams, yields a predicted lateral load of 1032 kips. This analysis places the predicted and test lateral load values within acceptable agreement. The comparison further substantiates the assumption that three-dimensional effects were significant during the full-scale tests.

Strength and Ductility

The comparison between medium-scale wall-frame specimen and full-scale three-dimensional specimen test results indicates that a lower bound on the strength of a structure is arrived at by the individual analyses of planar frames. These are assumed to function together in the direction of load by the shearing transfer action of horizontal floor diaphragms. The current practice in design of neglecting out-of-plane, three dimensional effects is therefore conservative from a strength standpoint.

Structural walls of all specimens tested exhibited ductile flexural behavior during hinge formation and beyond. Shear stress in the walls was moderate. For these structures, the increased strength of the system due to three-dimensional effects also means that the structural wall will be required to absorb higher shear stresses. However, increased vertical load on the walls increases ductility above that determined from a planar analysis and ignoring vertical loads.

CONCLUSIONS

When correlated with results of tests on a full-scale structure tested in Japan, results of tests of planar specimens suggested a lower bound to strength can be calculated by analyses of individual planar frames. The current practice in design of neglecting out-of-plane, three-dimensional effects is therefore conservative from a strength strength standpoint. Increased strength of the system due to three-dimensional effects means that the structural wall will be required to absorb higher shear stresses. However, vertical forces on the wall increase strength and rotational capacity above values determined from a planar analysis. The structural walls of all specimens showed ductile flexural behavior during hinge formation and beyond.

ACKNOWLEDGMENTS

The research was sponsored by National Science Foundation (NSF), under Grant No. PFR-8008753. Testing parameters and specimen designs were developed under the direction of the United States members of the Joint Technical Coordinating Committee of the U.S.-Japan Cooperative Earthquake Engineering Research Program.

REFERENCES

1. MORGAN, B. J., HIRAISHI, H., and CORLEY, W. G., "U.S-Japan Tests of Reinforced Concrete Structures, Comparison of Analysis and Test Results," Proceedings of the Fourth Meeting of the Joint Technical Coordinating Committee, U.S.-Japan Cooperative Research Program Utilizing Large-Scale Testing Facilities, Building Research Institute, Tsukuba-gun, Ibaraki-ken, Japan, June 16, 1983.

2. U.S.-Japan Planning Group, Cooperative Research Program Utilizing Large-Scale Testing Facilities, "Recommendations for a U.S.-Japan Cooperative Research Program Utilizing Large-Scale Testing Facilities," Report No. UCB/EERC-79/26, Earthquake Engineering Research Center, University of California, Berkeley, California, September 1979.

3. "Standard Specification for Deformed and Plain Billet-Steel Bars for Concrete Reinforcement," A615, American Society for Testing and Materials, Philadelphia, Pennsylvania.

DISCUSSION

Dr F K Garas   Taylor Woodrow Construction Ltd   UK

Having tested 1/3 scaled models, could you have achieved the same objectives by testing much smaller scaled models? This probably would have resulted in speed of construction in testing and also reduction in cost.

Author's Reply

In my opinion, tests of specimens having a scale in the range of 1.3 to 1.5 were necessary in the US/Japan full scale test program. Smaller scale structures were used and did give overall global response data. However, the larger specimens provided a more reliable representation of interelastic behaviour observed in the full scale structure.

It is my opinion that the most cost effective way of obtaining necessary information for earthquake response is to use specimens smaller than about 1.5 scale on shake tables, structures in the range of 1.5 to 1.4 scale under predetermined static loading, and full scale component test where information on behaviour of joints is required. All three types of test specimens have their unique advantages.

Finally, it is my opinion that occasionally full scale testing should also be done. Although the cost should be in the range of 50 times that of a medium scale assembly, significant information was found in the full scale structure that was not obtained from the smaller test specimens.

13 Dynamic response of models and concrete frame structures

DANIEL P. ABRAMS
University of Colorado, USA

SUMMARY

This paper presents final results of a study that was done to examine the use of reduced-scale models in predicting dynamic response of concrete frame structures. The study consisted of an experimental phase where hysteretic behavior of large-scale frame components were determined as well as that for models constructed at one-twelfth scale and one-quarter scale. An analytical study using nonlinear dynamic analyses followed where correlations were studied for response of large and small-scale structures.

INTRODUCTION

Models of multistory frame structures as small as one-twelfth scale have been used in many research laboratories to examine response of concrete buildings subjected to strong ground motions. Small-scale test structures have been fabricated using mortar and wire for modelling materials. In order for result of these investigations to be used for the design of full-scale structures, on must ask whether a structure composed of these materials may constitute the resistance mechanisms manifest in an actual reinforced concrete structure. The study described in this paper attempts to answer this question. Because it is impossible at present to test a full-scale multistory building on an earthquake simulator, and pseudodynamic test methods for this size of structu have proven to be very expensive, the study relied on experimental data from tests of components only. Load-deflection relations for beam-column assembla and base-story column members were compared for specimens constructed at a large scale and at approximately one-twelfth scale. Specimens were subjected to similar patterns of slowly applied displacement reversals. Sensitivities of building response to differences in hysteresis relations of components were evaluated using a nonlinear dynamic analyses. Computed response was normalized so that motion within the linear range would be equal. Limitations of the study were that (a) only planar behavior of frames and walls were studied, and (b) strain rate effects were neglected.

MEASURED SCALE RELATIONS FOR BEAM-COLUMN ASSEMBLAGES

A total of nine small-scale (1) and six large-scale (2) beam-column assemblages were tested. Specimens were indentical in configuration and relative amounts of reinforcement, and were subjected to essentially the same patterns of joint rotation. Imposed distortion histories were patterned from representative excursions which were established from shaking-table test data. Dimensions of the small-scale specimens were one-ninth of those for the large-scale specimens. Beams were designed relative to the columns so that inelastic behavior would occur in the beams at the faces of the columns. Shear stresses were low to insure behavior predominantly in flexure.

Reinforcing ratios in beams were approximately 1.0 percent. Small-scale reinforcement consisted of No. 13 gage (0.092 inch, 2.3 mm. diameter) bright basic wire which was annealed and redrawn to result in a yield stress of approximately 60 ksi (413 MPa). The surface of the wire was smooth. Measured stress-strain relations for the model reinforcement were bilinear with a well defined yield point for monotonically increasing stress.

Small-scale exterior beam-column assemblages (Fig. 1a) responded within linear and nonlinear ranges of response with nearly the same load-deflection relations as that measured for large-scale specimens. Longitudinal beam reinforcement was anchored sufficiently within the joint for each specimen. Although the large-scale specimens responded with a more uniform distribution of flexural cracks than that observed for the small-scale specimens, overall flexural stiffnesses and strengths were represented well at one-twelfth scale.

Tests of interior beam-column assemblages (Fig. 1b) did not show the good correlations in scale as observed for the exterior joints. For the interior joints, concentrations of bond stress within the joint resulted in local slippage of reinforcement. This action occurred for both large and small-scale specimens, but to varying extents. Bond was lost completely for the small-scale specimen across the width of the column during the first large-amplitude cycle of loading. A sharp reduction in stiffness within load reversal regions was a result of this slippage. After five large-amplitude cycles, the large-scale specimens showed partial bond deterioration and resembled the small-scale specimens during earlier cycles. Apart from this deviation, the small-scale specimens were able to mimic both the strength and stiffness characteristics of the large-scale specimens.

MEASURED SCALE RELATIONS FOR BASE-STORY COLUMNS

Typically in frame-wall structures, formation of an inelastic hinge is required at the base of the columns so that an energy dissipating mechanism may be developed. If this element cannot respond within tolerable limits of inelastic deformation, then the resistance of the overall structure will be reduced significantly. Because of the importance of this element on building response, tests were done on large and small-scale column members to evaluate scale relations. A total of eight small-scale (3) and ten large-scale (4,5,6) column specimens have been tested. Axial compressive forces were varied with lateral displacements to simulate forces on columns adjacent to the exterior bay. Modelling materials were the same as those used for the beam-column specimens. Tests of one-twelfth scale and one-half scale specimens revealed similar hysteretic relations despite size effects. The influence of the varying axial force on the opening and closing of flexural cracks was modelled well. Because of the good correlations, differences in behavior attributable to scale effect of these elements were not included in the analytical study which follows.

COMPUTED RESPONSE OF LARGE AND SMALL-SCALE BUILDINGS

A computational model was developed (7) to determine response of structures consisting of components at each scale. A time-step integration routine was written that considered acceleration to vary linearly across a time step. The structure considered for computations (Fig. 2) was a ten-story frame-wall structure. Behavior of both interior and exterior beam-column assemblages were considered. The base motion was patterned after the 1940 El Centro Earthquake (NS Component). Calculations were made for those forms of response measured during a test of a small-scale model: displacement accelerations and forces resisted by an internal wall (Fig. 3). Responses were normalized with respect to time and length factors such that response within the early stages, or linear range, would be coincident. Differences in response during subsequent periods were attributable to differences in modelling hysteretic characteristics for large and small-scale interior and exterior beam-column assemblages.

Conclusions that can be made from comparison of the response of a reduced-scale model and a full-scale concrete building are noted.
(a) Apparent natural frequencies of the small-scale structure were slightly less than those of the large-scale structure. Natural frequencies of both structures decreased with successive cycles of motion, however, the small-scale structure showed more rapid deterioration than did the large-scale structure.
(b) Maximum displacement and acceleration response occurred during the same cycle for both structures. Maximum displacements were nearly the same for each structure. Maximum accelerations were larger for the large-scale structure.
(3) Frequency contents of displacement and acceleration response were similar for both structures. Modal frequencies and base-motion frequencies were observed in records with similar participations for each structure.
(d) Amplitudes of lateral force resisted by either the wall or the frames were markedly different for each structure. Sequences and frequencies of wall response, however, were similar for each structure.

The reason for these tendencies was because of the deterioration in bond of beam reinforcement across the width of the column for the interior-joint specimen. Despite the fair correlation in moment-rotation relationships for the beams of the exterior-joint specimen (Fig. 4a), the moment-rotation relationship for the small-scale interior-joint specimen (Fig. 4b) revealed a significant softening within the load-reversal region after the first large-amplitude cycle of response.

Despite this difference, displaced shapes (Fig. 5) for each structure were nearly the same because displacements were relatively insensitive to shifts in frequency for this particular range and earthquake. Distributions of lateral force resisted by the overall structure were similar for both structures within early stages of the earthquake, but varied in later stages. Forces resisted by individual lateral-load resisting units did vary substantially for each structure during the entire earthquake. This was because the wall was modelled as linear for both structures which reflected differences in modelling of frame components.

Story shears resisted by the overall structure (Fig. 6) were similar in frequency content, sequence and amplitude for both structures. Participation of the wall in resisting these shears, however, varied for each structure at the upper levels. At the lower levels, similar wall shears were seen for each structure.

CONCLUDING REMARKS

Results of the study have shown that a model as small as one-twelfth scale can be used to capture general characteristics of a reinforced concrete building vibrating within the nonlinear range of response. Overall response of a ten-story structure could be depicted with sufficient accuracy to allow for studies of response of the complete structural system. Forces resisted by individual frames or wall were sensitive to relative differences in hysteretic characteristics of large and small-scale interior beam-column assemblages. These differences were attributable to the difficulty in simulating the deterioration of bond with repeated and reversed displacements.

Further study should examine correlations in response for different base motions and structural configurations.

ACKNOWLEDGMENTS

The research was funded by the National Science Foundation of the United States (Grant No. CEE-8119385). Drs. John B. Scalzi and Michael Gaus are the cognizant program directs. Experimental work involving the large-scale specimens and the analytical study were done at the University of Colorado. Graduate research assistants J. Stewart, R. Bedell, H. Davis, P. Philleo, and W. Epp were responsible for testing. The computations were done by S. Tangkijngamvong using a Cyber 175 computer at the University of Colorado. Experimental work for the small-scale specimens was done at the University of Illinois under the direction of Professor M. A. Sozen, and the author.

REFERENCES

(1) KREGER, M E, and ABRAMS, D P, 'Measured Hysteresis Relationships for Small-Scale Beam-Column Joints', University of Illinois, Civil Engineering Studies, <u>Structural Research Series No. 453</u> (August 1978).

(2) PHILLEO, P, and ABRAMS, D P, 'Scale Relationships of Concrete Beam-Column Joints', University of Colorado, <u>Structural Research Series No. 8301</u> (December 1983).

(3) GILBERTSEN, N D, and MOEHLE, J P, 'Experimental Study of Small-Scale Reinforced Concrete Columns Subjected to Axial and Shear Force Reversals', University of Illinois, Civil Engineering Studies, <u>Structural Research Series No. 481</u> (July 1980).

(4) BEDELL, R, and ABRAMS, D P, 'Scale Relationships for Concrete Columns', University of Colorado, Structural Research Series No. 8302 (January 1983).

(5) DAVIS, H, 'Behavior of Columns at the Base of Concrete Frame Structures', M.S. Thesis, Department of Civil, Environmental, and Architectural Engineering, University of Colorado at Boulder (August 1983).

(6) EPP, W H, 'The Effect of Loading Path on the Hysteretic Response of Concrete Columns', Department of Civil, Environmental, and Architectural Engineering, University of Colorado at Boulder (July 1984).

(7) ABRAMS, D P, and TANGKIJNGAMVONG, 'Dynamic Response of Reduced Scale Models and Reinforced Concrete Structures', University of Colorado, <u>Structural Research Series No. 8401</u> (February 1984).

Figure 1 Measured Load-Deflection Relations for Beam-Column Assemblages

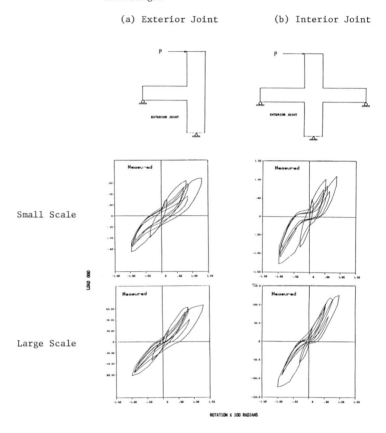

Figure 2 Structure Considered for Computations

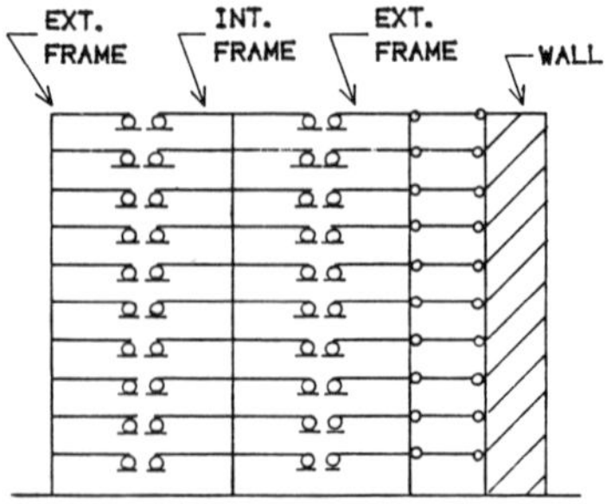

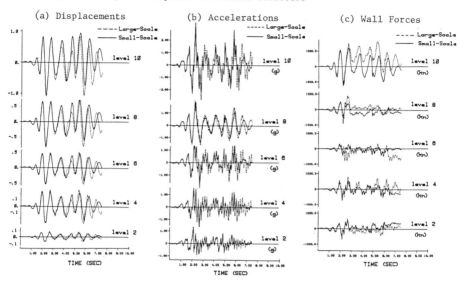

Figure 3   Computed Response of Overall Structure
(a) Displacements   (b) Accelerations   (c) Wall Forces

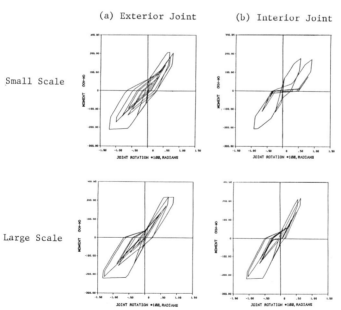

Figure 4   Computed Moment-Rotation Relations for Beams
(a) Exterior Joint   (b) Interior Joint

Small Scale

Large Scale

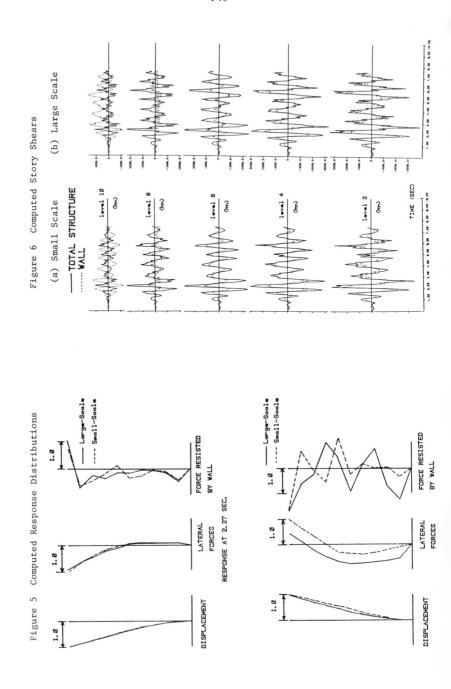

Figure 5 Computed Response Distributions

Figure 6 Computed Story Shears

(a) Small Scale (b) Large Scale

Author's Closure

In response to Dr Fumagalli's comments regarding the use of physical and numerical models for earthquake research, it may be worthwhile to state the purpose of the investigation described in the paper and the chronology of experiments. Rather than model a prototype structure at a reduced scale, the research program did the reverse. Components of model frame structures were expanded to a large scale and tested under slowly applied reversals of lateral deflection to verify the accuracy of the models in representing hysteretic behaviour. A numerical model was used solely to examine sensitivities of dynamic response to scaling relations of components. The purpose of the study was to verify past studies using reduced-scale models of concrete structures subjected to simulated earthquake motions.

In response to Professor Wright's question regarding the unrealistic simplicity of the large-scale component configuration, it must be stated that a direct correlation was intended with the small-scale components. Because the small-scale test structures were idealized models of lateral-load resistance, and not replicas of actual prototype structures, the configuration of the large-scale components was slightly abstract. However, it was possible to identify resistance mechanisms under repeated and reversed loading which could be correlated with more realistic large-scale components tested by others.

In response to Dr Somerville's discussion regarding the general use of models for predicting response to a prescribed load, I would like to comment on scaling correlations for large-amplitude non-linear dynamic problems in reinforced concrete members that are controlled by flexure. For these members, the crucial parameters that must be simulated are the stress-strain relations for the reinforcement, particularly at a relatively high strain rate, and the bond-slip behaviour under repeated and reversed deflections.

Tests of model reinforcement at high strain rates have been done at the University of Illinois (1). These tests have shown that the separation between upper and lower yield points is influenced by strain rate, and that knurling of the wire will result in a marked reduction in this separation. The limitation of this test series, however, is that strains were increased monotonically to failure. A test structure oscillating through a number of large-amplitude cycles would experience this strain history only once. The remaining large-amplitude cycles would not contain a discrete yield point but would be governed by the Bauschinger effect. The importance of strain rate on this effect has not been studied for model reinforcement.

Anchorage of small-scale wire in model concrete has also been examined at the University of Illinois (2). Tests have shown that a wide scatter in bond strength can occur for both knurled and unknurled bars, but a safe value of bond strength can be established and used in the design of a model structure. It should be noted, however, that it is still nearly impossible to simulate the deterioration in bond stiffness, particularly for an interior beam-column joint, with repeated and reversed deformations despite the scale. For this reason, care should be taken when interpreting measured response of a model. Further research needs to be done to examine this deterioration and develop new methods to improve bond characteristics for both large and small-scale reinforcement.

REFERENCES

(1) Staffier, S R, and M A Sozen, "Effect of Strain Rate on Yield Stress of Model Reinforcement", Civil Engineering Studies, Structural Research Series No 415, University of Illinois, Urbana, February 1975.

(2) Gavin, N L, "Bond Characteristics of Model Reinforcement", Civil Engineering Studies, Structural Research Series No 427, University of Illinois, Urbana, April 1976.

14 Impact testing on microconcrete models

W. THOMA and R. K. MÜLLER
Institute for Structural Model Analysis, University of Stuttgart, FRG

SUMMARY

In order to employ the advantages offered by microconcrete models also for impact testing, different beam models were used to demonstrate that reproducible results may be obtained even with relatively small scales; they essentially correspond to tests carried out on prototype beams.

1. INTRODUCTION

At the Institute for Structural Model Analysis, University of Stuttgart, a technique for producing small-scale concrete models has been developed which makes it possible to determine under realistic conditions the cracking- and failure loads of reinforced concrete structures as well as the cracking distribution (1). Within the framework of an experimental programme sponsored by the "Deutsche Forschungsgemeinschaft", the task set was to examine whether the microconcrete technology developed was equally suitable for conducting impact tests with small-scale models. It had to be demonstrated that the test results obtained with models were also relevant in the case of dynamic loading with prototype structures.

For comparison the impact tests on reinforced concrete beams and plates carried out by Eibl, Dortmund, were used (2), (3), (4), (5). The following will show some examples of the results of the beam tests which may be of interest, despite the fact that several other impact tests on beam models have already been conducted by different research workers. At present, the plate experiments are not yet concluded.

2. APPROACH TO THE MODEL DESIGN

With regard to the model laws (and the measurement values) strict similarity was aimed at, that means:

$$\sigma_v = \frac{\sigma \text{ model}}{\sigma \text{ prototype}} = E_v = \tau_v = 1.$$

Furthermore, the following known relations apply:

$$\varepsilon_v = 1$$
$$\rho_v = 1$$
$$F_v = E_v \, L_v^2 \ .$$

The mixing formulations for the microconcretes and the treatment of the wires were adapted to the actual strength of the prototype. For this, the production methods presented by Sautner (6) were developed further.

Table 1 Example showing the similarity of the characteristic values of the material (mean values)

T - BEAMS B 81/4 REINFORCEMENT MATERIAL	PROTOTYPE BSt 420/500RK	MODEL-SCALE	MODEL T - BEAMS 1 - 4 ST 37K-TEMPERATURE TREATED -PROFILED		USE AS
			NOMINAL	ACTUAL VALUE	
1) Ø 12 mm					
-ACTUAL DIAMETER (mm)	12,17	L_v	1,43	1,42	
$-\beta_{0,2}$ (N/mm^2)	511	1	511	535	STIRRUP
$-\beta_z$ (N/mm^2)	610	1	610	593	REINFORCE-
$-\delta_{10}$ (%)	9,5	1	9,5	-	MENT
2) Ø 16 mm					
-ACTUAL DIAMETER (mm)	16,07	L_v	1,89	1,87	COMPRESSION
$-\beta_{0,2}$ (N/mm^2)	435	1	435	487	REINFORCE- MENT
$-\beta_z$ (N/mm^2)	519	1	519	552	
$-\delta_{10}$ (%)	12,8	1	12,8	7,8	
3) Ø 22 mm					
-ACTUAL DIAMETER (mm)	22,22	L_v	2,61	2,60	
$-\beta_{0,2}$ (N/mm^2)	471	1	471	504	TENSION
$-\beta_z$ (N/mm^2)	597	1	597	559	REINFORCE-
$-\delta_{10}$ (%)	13,7	1	13,7	8,0	MENT
4) Ø 28 mm					
-ACTUAL DIAMETER (mm)	28,20	L_v	3,32	3,32	
$-\beta_{0,2}$ (N/mm^2)	402	1	402	405	TENSION
$-\beta_z$ (N/mm^2)	492	1	492	444	REINFORCE-
$-\delta_{10}$ (%)	15,3	1	15,3	9,7	MENT
CONCRETE					
-MAX. GRAIN (mm)	16	(L_v)	(1,9)	2,0	
-COMPRESSIVE CUBE STRENGTH (N/mm^2)	40,8	1	40,8	42,9	
-BENDING TENSION STRENGTH (N/mm^2)	4,5	1	4,5	4,5	
-TENSILE SPLITTING STRENGTH (N/mm^2)	3,4	1	3,4	3,9	
-YOUNG'S MODULUS (N/mm^2)	29450	1	29450	29130	

As shown by table 1, the greatest difference of material occurs at the failure elongation δ_{10} of the reinforcement. If required for the microconcrete reinforcement, the failure elongation as well as the ratio failure strength/yield point may be considerably increased by treating the wire once more at a certain temperature subsequent to profiling. For beam tests, however, the amount of time and effort involved at present does not appear to justify this procedure.

In addition to the model laws for static loading, the laws for dynamic loading must also be observed, that means:

$$\frac{E_v}{\rho_v} = \frac{L_v^2}{t_v^2} \quad \text{and also} \quad v_v = \frac{L_v}{t_v}.$$

For the tests carried out by Eibl, no measurement results on the rate of velocity of the impact waves were available.

During pre-testing with different microconcretes having a hardened concrete gross density of approx. 2100 kg/m³, a relation similar to normal-weight concrete could be recognized between the concrete strength and the velocity of the longitudinal waves v_L, so that for the time analogy of the measurement signals sufficient agreement could be expected.

As a freely chosen model parameter the length scale L_v was employed.

3. EXTENT OF THE TESTS

Altogether 12 microconcrete models with varying length scales were tested (scales 1:5, 1:7,8 and 1:8,5). A hard impact with a non-deformable impact mass was applied, so that even in the case of relatively low impact velocities ($v_0 = 0,7$ m/s - 13,2 m/s) the test specimens were subjected to dynamic behaviour with the following characteristics:
- the greatest part of the kinetic energy is absorbed by the beam exposed to the impact
- very short impact times (a few milliseconds)
- distinct maximal peak of impact power within less than a millisecond after start of contact.

3 different types of beams were selected from Eibl's experimental programme. Differing types of failure served as criteria for selection (see Fig. 2).

Figure 2 Selected significant details of the impact tests

	PENDULUM TYPE IMPACT TESTING MACHINE	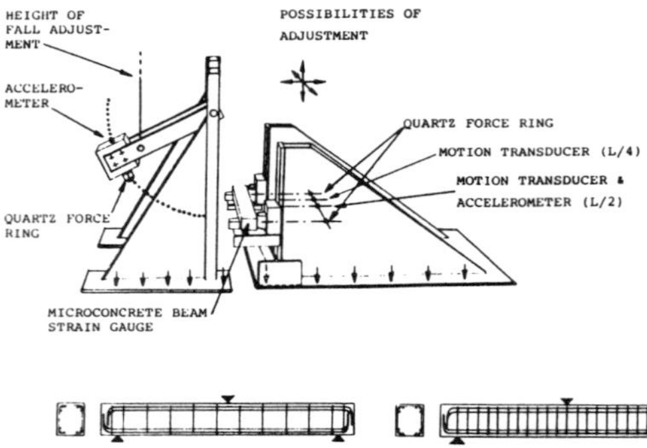	FALLING WEIGHT & RELEASING DEVICE	
TEST NR.	3.6	3.7		B 81/4
RATE OF FALL	0,7 - 2,5 m/s	0,7 - 2,65 m/s		13,2 m/s
SCALE	1:5	1:7,8	1:5	1:7,8	1:8,5
QUANTITY				
-PROTOTYPE		1	1	1
-MODEL	2	2	2	2	
FAILURE	SHEARING STRAIN	BENDING STRAIN	3 DYNAMIC LOAD ; 1 STATIC LOAD ROTATION (L/2)	BENDING STRAIN
IMPORTANT FACT OF TESTING	REPRODUCIBILITY OF TESTING RESULTS AT DIFFERENT SCALES SIDE OF CONCRETE		SIDE OF REINFORCEMENT	

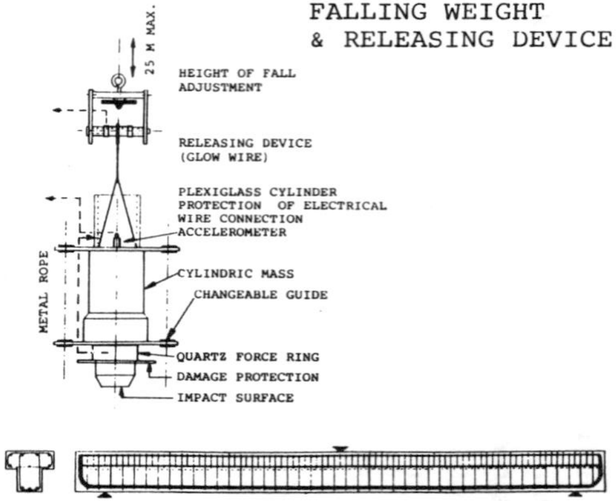

The tests 3.6 and 3.7 are in actual fact transverse impacts on columns. However, since these were tested in a lying position by pendulum impact, they are, for all practical purposes, to be viewed as beams with rectangular cross sections. Here, no. 3.6 and 3.7 only differ as far as the stirrup distance is concerned. In the case of the T-beam test B 81/4 the falling body was additionally accelerated for the original by a compressed-air device which was not necessary for the models since a suitable shaft was available for the impact tests where speeds up to 22 m/s could be achieved in a free fall.

Just as for the prototype beams, the following measurement values were registered during the model tests:

- for the impact mass:
  - impact force
  - acceleration
  - rate of fall
- for the beam:
  - steel strain
  - deflection
  - acceleration
  - supporting force.

Some of the measurement values are presented in sections 4 and 5. They have been converted according to the model parameters given in section 2, to the scale of the prototypes.

4. BEAMS WITH RECTANGULAR CROSS SECTION

For this series of beams different concrete mixes were produced. Riversands were used as aggregate. The grading curve was selected on the basis of a modified Fuller's parabola and divided into 4 grain fractions, the cement (PZ 45 F) serving as the smallest grain fraction. For the 1:5 scale a maximum grain of 4 mm diameter, and for the 1:7,8 scale a maximum grain of 2 mm diameter were used. In order to obtain a ratio of bending tension/ compression strength affined to the prototype beams, a water/cement value of 0,50 was chosen. Since, however, the concrete strengths of the prototype beams were relatively low (β_w = 24,8 N/mm^2; β_{BZ} = 2,3 N/mm^2; Young's modulus = 26.200 N/mm^2), the aggregates were coated with a silicone resin agent which reduces the bonding tensile strength and, furthermore, small amounts of a silicone resin solution were mixed into the wet cement in order to produce slight disturbances in the hardened cement stone.

Figure 3 Example of the similarity of the deflections at $(\frac{L}{2})$ Beam 3.6

Figure 4 Example of the similarity of the steel strain in the tensile zone at (0,19 L)   Beam 3.6

Figure 5 Example of the similarity of the support forces (left support)   Beam 3.7

Figure 6 Example of the similarity of the steel strains in the compression zone at (0,4 L)   Beam 3.7

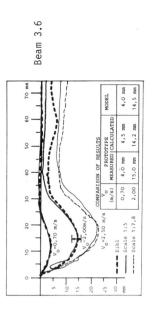
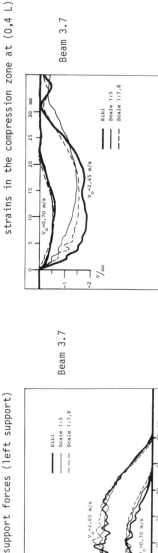

As shown by Fig. 3-6, the model results for the different rates of fall are in good agreement of amplitude and time period with the measurements carried out by Eibl. Fig. 3 additionally contains a table presenting measurements vs calculation.

Since each time 2 beams of the same scale were tested under identical conditions, the results are understood to be reproducible. The measurement results of repeat tests essentially differed only as far as the steel strain was concerned which - as is known - depends on the accidental distance of the strain gauge to the crack.

In the case of the force measurements with similar beams, the greatest deviations occurred with the amplitude of the first peak.

This applied to the prototype beams as well as to the models. The measurements of the left and right supporting forces showed the following standard deviations (related to the resp. mean value):
- for the prototype beams: $v = 7,4 \% - 23,0 \%$ (n = 2)
- for the models: $v = 7,3 \% - 35,8 \%$ (n = 2 to 9).

By comparison, the forces in the medium range of the contact period showed a good reproducibility:
- for the prototype beams: $v = 3,0 \%$ (n = 2)
- for the models: $v = 5,3 \%$ (n = 2 to 9).

n ... number of measurement values per mean value
v ... coefficient of variation

5. T-BEAMS

The T-beam selected was part of an analysis of different types of reinforcement and was designed by Eibl for the clear failure of the lower longitudinal reinforcement. The test specimens were exposed to impact loading far into the plasticity range. The model tests were to demonstrate whether a similarity can still be achieved within the limit range of the plastic load.

The 4 model beams were produced with an identical microconcrete mix and without the use of silicone resin. The water/cement value amounted to 0,60.

5.1 Static Loading

Under otherwise identical conditions, a model beam was statically loaded up to failure of the tensile reinforcement. Here, the maximal bending moment obtained was M_B = 251,1 KNm as opposed to the prototype beam where the calculated maximal bending moment was M_B = 256 KNm.

5.2 Dynamic Loading

Corresponding to the prototype beam, the supports were prestressed in order to prevent the lifting of the T-beam at the beginning of the impact. As shown by Fig. 7, the cracks in the model occur in the plate area near the support as well as at the point of impact. As far as the crack distances are concerned, there is also good agreement between the model and the prototype beams. However, since the failure elongation δ_{10} of the model tensile reinforcement was approx. 60 % smaller than that of the prototype beam (see Table 1), and since also the ultimate shear stress was somewhat less, the model beam, by consequence, has a lower rotational capability. Contrary to the prototype beam, a failure of the lower tensile reinforcement occurred, 36 to 46 ms after start of impact (converted to prototype scale). With the prototype beam the maximum deflection occurred after 44,5 ms. Thus, the microconcrete beams only failed shortly before the maximal structural deflection, at a time where the loading action was almost completed. Therefore, the impact force signals do not only display good agreement for the first and second stage of the loading action (presented in Fig. 8), but also for the third stage up to failure. At present, work is in progress to find

Figure 7   Example showing the similarity of the cracks

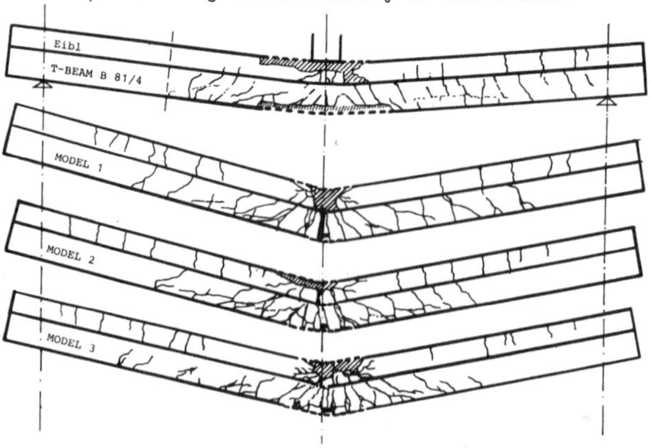

a feasible solution for increasing the strain capability of microconcrete models since further results have demonstrated that microconcrete may very well be employed for analyses with extreme material loading.

Figure 8 Example showing the similarity of the impact forces
T-beams B 81/4

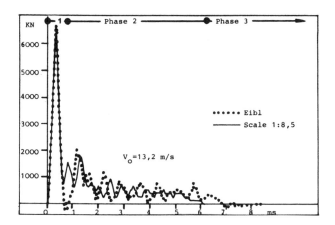

REFERENCES

(1) MÜLLER, R K, 'Microconcrete for Structural Model Analysis', Seminar Report, Institute for Structural Model Analysis, University of Stuttgart.

(2) EIBL, J and BLOCK, K, 'Zur Beanspruchung von Balken und Stützen bei hartem Stoß (impact)', Bauingenieur no. 56 (October 1981).

(3) EIBL, J and BLOCK, K, 'Stoßbeanspruchung von Stützen durch Fahrzeuge und andere Stoßlasten, Final Report AZ: B II 5-81 0705-213/1, University of Dortmund (1980).

(4) EIBL, J, BLOCK, K and KREUSER, K, 'Vergleichende Versuche an stoßbelasteten Balken und Platten mit Bewehrung aus Betonstahl 1100 bzw. herkömmlichen Betonstählen', Research Report AZ: 03 FKH 201, University of Karlsruhe (October 1983).

(5) EIBL, J and KREUSER, K, 'Durchstanzfestigkeit von Stahlbetonplatten unter dynamischer Beanspruchung', Research Report AZ: B I 7-81 0705-258, University of Dortmund (1982).

(6) SAUTNER, M, 'Ein Beitrag zur Entwicklung der Mikrobetontechnik', Reports of the Institute for Structural Model Analysis, vol. no. 7, University of Stuttgart (1983).

15 Impact resistance of reinforced concrete slabs

ALAN J. WATSON and THAMIR K. AL-AZAWI
Department of Civil and Structural Engineering, University of Sheffield, UK

SUMMARY

Simply supported one-way spanning model scale reinforced concrete slabs were subjected to both impact loading from a falling mass and static loading. The impact force, reaction-time histories and the transient deflections were recorded for each impact. The independent variables were the impact test conditions and the slab thickness. After each impact test the morphology and the residual width of the cracks, the residual deflection and the residual static load resistance of the slab were measured. Test results were compared with results from similar impact experiments done on larger concrete slabs [1].

INTRODUCTION

Most structures are designed to withstand only static loads but some are designed to also resist dynamic loads such as those due to wind, earthquake and impact. The impact loads are often accidental and important questions about probability and consequence need to be answered before deciding whether to design for impact resistance. At present little reliable data for impact design are available and the design process is complicated. In these circumstances, investigation of the inherent impact resistance of structures designed for static loads, can be of assistance to structural engineers who are faced with the decision of whether or not, a full impact design needs to be attempted.

This paper describes the investigation carried out on one-way spanning simply supported reinforced concrete slabs which are of the same material, dimensional ratios and reinforcement, type and percentage, typically found in conventional reinforced concrete construction. For practical convenience the slabs were constructed at model scale and scaling laws applied so that these results could be compared with those reported by Fujii and Miyamoto (1) (on larger slabs hereafter called the prototype).

Careful attempts were made to satisfy the similarity conditions between model and prototype slabs.

EXPERIMENTATION

Test Specimens

The model specimens were one-way spanning simply supported slabs of dimensions 525mm x 465mm having a span length of 500mm and a thickness of either 54mm or 42mm.

These slabs are intended to represent reactor containments which are normally between 0.5m and 2.0m thick, so that the model slabs are at a scale between 1:10 and 1:40.

Materials and Fabrication of Specimens

The concrete used for the models had 10mm max. size aggregate, ferrocrete-rapid hardening portland cement and zone 2, BS 882 river sand with a max. particle size BS sieve No. 7. The ratios by weight of (total agg./cement), (sand/total agg.) and (water/cement) were 4.0, 0.55 and 0.61 respectively. The mean crushing strength of 6 No. 50mm dia. x 100mm concrete cylinders was 29 N/mm^2 at 14 days and the splitting tensile strength was 3.4 N/mm^2.

The reinforcement was deformed high-yield steel bars having a yield stress of 590 N/mm^2 and ultimate strength of 685 N/mm^2. Fig. 1 shows the reinforcement details.

Figure 1 Reinforcement details

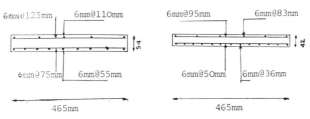

The slabs were demoulded 24 hrs after being cast and then kept for 6 days in the curing room at 22 ± 0.5°C, 95% - 99% RH. Two days before the testing date (at age 14 days) the slabs were painted white so that cracks would be more clearly seen.

Test Equipment and Procedure

A test rig was designed so that it could be used for both the falling mass impact and the static tests, Fig. 2. This meant that the slab support conditions were always the same. The impact or static load was applied at the centre of the slab through a solid high strength steel bar 50mm dia. x 500mm long hereafter called the pressure bar. A 6mm thick x 62mm square loading steel plate was placed between the pressure bar and the concrete slab.

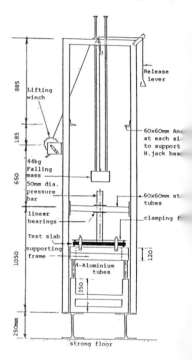

Figure 2 The test rig

Slab deflections at four points were measured by linear variable differential transformers LVDT1 to LVDT 4, Fig. 3. Electrical resistance strain gauges, (ERS), on the pressure bar, P.L, were used to measure the applied load and the slab reactions were measured by load cells using ERS gauges on aluminium tubes, RL1 and RL3 respectively (Fig. 3). ERS gauges were also bonded to the reinforcement under the impact point. The gauges were Japanese foil electrical resistance strain gauges, FKL, 120Ω, 5mm gauge length and were connected into a Wheatstone bridge circuit so that the bending strains were cancelled.

In the static tests a hydraulic jack was used and at each 2kN load increment the deflections and reinforcement stresses were recorded and all visible cracks were marked. This procedure was continued until failure.

In the impact tests a 44kg steel mass was used and the drop height was varied from 25mm to 1200mm. The transient measurements at each strain gauge point and deflections from the LVDT's were recorded during each impact.

For each specimen type three identical slabs were tested, each under different impact conditions as follows:

1. Regularly increased severity of impact similar to that carried out on the prototype slabs. The first impact was from a drop of 25mm. The drop was then increased in steps of 25mm until the appearance of the first visible crack. The drop was then increased in steps of 50mm until failure occurred. Failure was deemed to be when a cone of concrete was disloged under the impact point. A cycle of impacts was carried out to each new maximum drop. For instance in milimetres drop; 1st cycle: 25, 50, 25. 2nd cycle: 25, 50, 75, 50, 25. 3rd cycle: 25, 50, 75, 100, 75, 50, 25 etc. This was the only impact test condition followed in the prototype investigation and presumably was adopted in order to more closely identify a critical drop height.

2. A single impact blow from a drop height of $H = A/W_s$ where A denotes the area under the static load - deflection curve for a slab and W_s the weight of the falling mass. In these experiments the impact energy was equal to the static energy absorption capacity.

3. A single impact blow from a drop height of $H_\alpha = H/\alpha$ where α denotes a Simm's factor which allows for the slab inertia (3), and $\alpha = (1+0.8\ M_b/M_s)^{-1}$ where M_b and M_s are the slab mass and the falling mass respectively.

Figure 3 Diagram of test equipment

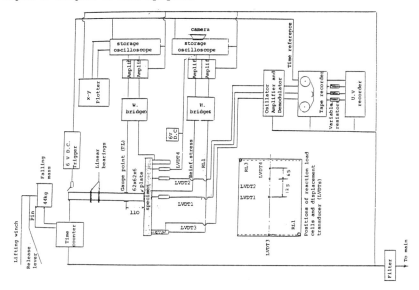

TEST RESULTS

For slabs subjected to a static load or an impact load of regularly increased severity, the initial visible cracks occurred at the slab centre and were mainly transverse to the direction of span in the static tests but were irregular in the impact tests and radiated from the impact point. As the static load or impact drop height was increased, existing cracks spread and new cracks were developed.

At failure, punching shear occurred and a cone of concrete was disloged beneath the load point, Fig. 4.

Figure 4   Slab failure modes

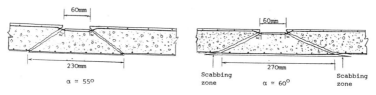

None of the slabs failed when they were subjected to just one impact blow and f the same slab thickness that subjected to an impact blow from a drop height of H_α had larger residual crack widths and deflection than that impacted from the smaller drop height, H, Fig. 5, Table 1. Typical transient records for slab D1 are shown in Fig. 6 (see Table 1 slab code for test conditions).

DISCUSSION

Impact Force-time History

The force-time history recorded at gauge point (PL) is the resultant of all reflected and transmitted stress waves through the system passing point PL. Accordingly the force-time history at the gauges in the pressure bar, PL (Fig. is not the same as that applied to the slab at the interface between the loadi plate under the pressure bar and the slab. The transient records shown in Fig indicate that the slab response is quasi-static because the reinforcement stre support reaction and slab deflection are all in phase and approximate a sinusoi shape. Therefore as a simplification, a sinusoidal impact force-time history with the same impulse and duration as that recorded at point (PL) was assumed.

Figure 5 Crack patterns (see Table 1 Slab code for test conditions)

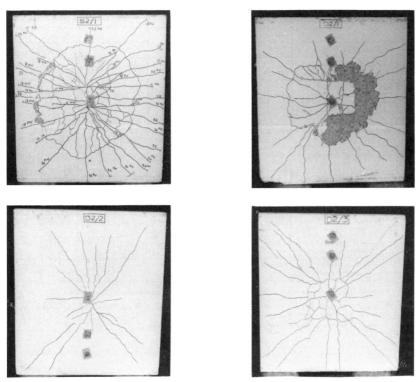

Figure 6 Transient records for slab D1/2

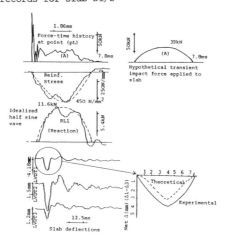

Table 1   Summary of test results

		Slabs 54mm thick				Slabs 42mm thick		
	S1	D1/1	D1/2	D1/3	S2	D2/1	D2/2	D2/3
Test condition**	S	R	H1	$H_\alpha 1$	S	R	H2	$H_\alpha 2$
1st visible load (kN)	16.9 (14.1)*	21.3	-	-	10.7 (12.0)	15.3 (12.0)	-	-
1st visible cracking height (mm)	-	150 (100)	620	1165	-	100 (83)	350	660
Failure height of drop (mm)	-	550 (380)	-	-	-	380 (248)	-	-
Failure mode+	P (P)	P, S (P, S)	No Failure	No Failure	P (P)	P, S (P, S)	No Failure	No Failure
Max. recorded load (kN)	45.0 (39.1)	37.0 (46.1)	39.0	64.0	38.0 (37.5)	27.5 (26.6)	32.1	39.6
Max. deflection (mm)	7.2	4.8 (3.9)	4.0	6.3	5.2 (4.3)	4.5 (3.8)	3.7	5.3
Max. reinforcement stress (N/mm^2)	561	-	450	558	247	-	184	220
After test — Max. residual crack width (Lower face)(mm)	1.00	2.00	0.10	0.40	0.13	0.80	0.02	0.04
After test — Max. residual deflection (mm)	8.80	3.8 (4.5)	0.90	2.10	4.10	4.4	0.61	0.86
After test — % reduction of static load resistance	failure	Failure	0.40	10.8	Failure	Failure	0.00	29.1

* Values in parantheses refer to prototype results scaled for equivalence to test slabs

** Where S denotes static; R regularly increased severity of impact; H one impact blow (H_1 = 620mm, H_2 = 350mm) and H_α one impact blow ($H_\alpha 1$ = 1165mm, $H_\alpha 2$ = 660mm)

\+ Where P denotes punching shear failure and S concrete scabbing

Following the theory of forced vibration for beams and assuming that the one-way spanning slab vibrates approximately as a beam, the central slab deflection was calculated using the method described by Hughes and Spiers (2) for the case of single impact based on the assumed sinusoidal transient impact force and using the moment of inertia of a cracked section.

Calculations indicated that the first mode of vibration predominated and the response is quasi-static which agrees with test results. The experimental and theoretical central deflection also show a good agreement, Fig. 6.

Comparison between Model and Prototype Slabs

A geometric linear scale factor of 0.417 exists between the model and prototype slabs and the falling mass was scaled-down by a factor of $(0.417)^3$ (4). The prototype reinforcement was (13mm) dia. deformed steel bar with a 340 N/mm^2 yi stress. The cylinder crushing strength, tensile splitting strength and Young' modulus of the prototype concrete were 28.1 N/mm^2, 3.0 N/mm^2 and 26.9 kN/mm^2

respectively as determined from 150mm dia. x 300mm long cylinders.

The prototype slabs were impact tested only under the scheme of regularly increased severity of impact and the available results are scaled-down for equivalence with the model slabs, Table 1, using the factors $(0.417)^2$, 0.417 and 1.0 for the load, deflection and height of drop respectively.

The comparison indicates that the height of drop at which the first visible crack and at which failure occurred are 17% - 33% higher in the model than in the protptype slabs. This may be attributed to more flexible supports for the model slab which would reduce the impact factor and also, to the higher splitting tensile strength of the model concrete. Comparison of the transient and residual deflections and the applied loads show good agreement with a maximum discrepancy of 19.5% in the maximum impact force on slab D1/1. The impact force applied to the prototype slabs was measured by an accelerometer attached to the hammer, and the impact force-time history was said to be sinusoidal as obtained in the model tests described above.

The reinforcement stress in the model and prototype was less than the yield stress and the crack pattern and failure modes were the same in both model and prototype slabs, Table 1.

Design Application

These experiments have shown that the model slabs did not have any loss of static load carrying capacity after being impacted by a single blow of energy equal to the static energy absorption capacity. Under normal circumstances this would be a very small impact, probably much below that representing realistic risk levels in most industrial structures. When the impact was increased by a factor of $\frac{H_\alpha}{H} = 1.88$ then there was a reduction of between 10% to 30% in the static load resistance and a considerable increase in the residual deflection and crack width. RC structures designed to the recommendation of CP 110 without considering impact loads, can be expected to have a safety factor of about 1.5 x 1.15 = 1.7 between the ultimate limit state and the serviceability limit state in flexure and this represents a safe loss of 40% in the static load resistance for structures damaged by impact or any other cause not specifically considered in the design. Similarly, for shear resistance, the safe loss would be about 55%.

Comparisons between model and prototype experiments were only made for the case of increased severity of impact but this has shown that model results can be scaled to predict prototype results.

CONCLUSIONS

1. Experiments on 0.417 scale model reinforced concrete slabs under repeated impact loads have shown that the scaled load at which the model slab cracked and that at which failure occurred, were 17% to 33% higher than for similar prototype slabs under a similar impact load.
2. Crack patterns and failure modes were the same for both model and prototype sl
3. There was no loss of static load carrying capacity for the model slabs after they had been impacted by a single blow of energy equal to the static energy absorption capacity of the slab.
4. The static load carrying capacity of the model slabs was reduced by 10% to when the impact energy was 88% greater than the static energy absorption capacity. Most RC structures should be able to lose about 40% of the static load strength before there was a danger that collapse of the damaged struct could occur under the service loads.

ACKNOWLEDGEMENTS

The authors are grateful to all the members of the staff of the Civil and Structural Engineering Department who have supported this research and assiste in the preparation of this paper. Acknowledgement is also given to the Govern of Iraq who are supporting Mr T K Al-Azawi.

REFERENCES

1. Manabu Fujii and Ayaho Miyamoto, "Improvement of the Impact Resistance for Prestressed Concrete Slabs". Inter-Association Symposium on Concrete Structures Under Impact and Impulsive Loading, Berlin, June 1982.

2. Hughes, G, Spiers, D M, "An Investigation of the Beam Impact Problem", C & Technical Report 546, April 1982.

3. Simms, L G, "Actual and Estimated Impact Resistance of Some Reinforced Concrete Units Failing in Bending", Journal of the Institution of Civil Engineers, No. 4, 1945.

4. Barr, P, Carter, P G, Howe, W D and Neilson, A J, " Replica Scaling studies of Hard Missiles Impacts on Reinforced Concrete". Inter-Association-Sympos: on Concrete Structures Under Impact and Impulsive Loading, Berlin, (June 1

Author's Closure

The equation used to calculate the central deflection of the slab under impact[2] was that for a one-way spanning slab assumed to be vibrating as a beam.

$$ie \quad Y(L/2,t) = \frac{2}{m_b} \sum_{i=1,3}^{\infty} \frac{1}{w_i} \int_0^t F(\bar{t}) \cdot \sin\left\{w_i(t - \bar{t})\right\} d\bar{t}$$

where

- y = Slab central deflection (mm)
- L = Span of slab (m)
- E = Concrete modulus of elasticity (N/m^2)
- I_c = Moment of inertia of a cracked section (m^4)
- m_b = Slab mass (kg)
- $F(t)$ = Impact force at time t. (N)
- t = Time (sec)
- $\bar{t}$ = Any time within the period (0 - t). (sec)
- i = Vibration mode number.
- w_i = Angular frequency of free vibration. (rad/sec)

where

$$w_i = i^2 \pi^2 \sqrt{\frac{E\,I_c}{m_b L^3}}$$

The above deflection equation was solved by numerical integration.

16 Experimental and numerical investigations of reinforced concrete structural members subjected to impact load

J. HERTER, E. LIMBERGER and K. BRANDES
Bundesanstalt für Materialprüfung (BAM), Berlin, FRG

SUMMARY

The safety analysis of nuclear power plants in regard to design laws in the Federal Republic of Germany includes the consideration of an aircraft impact on the reinforced concrete containment. To close gaps of knowlegde on the behaviour of concrete structures subjected to impact load, tests on RC-beams, RC-slabs, and reinforcing steel as well as numerical investigations have been performed. The specimens in beam and slab tests may be regarded as a cut-out of a RC-structure in an appropriate scale and were fabricated like regular RC-structural members and tested in a servohydraulic testing set-up. The intention in the tests was to obtain the same type of phenomena as occurs in many real RC-structures. We investigated the deformation rate effect on plastic hinges or yield lines and on the punching shear behaviour by performing static and impact tests on similar specimens. The strain rate influence of different grades of reinforcing steel has been investigated in tension tests. First results show that an influence of the strain rate or deformation rate on the stress-strain curve or the load-deformation curve exists which has to be taken into account in a design model. Associated numerical investigations were undertaken by using a Finite Element Model. The interaction between design requirements, concept of tests, formulation of reliable design models and analysis of the test results is shown.

INTRODUCTION

The safety analysis of nuclear plant facilities according to German design codes has to take into account an aircraft impact on the containment structure. In the early seventies, the knowledge of the behaviour of concrete structures subjected to this "load case" was not sufficient for an economic and realistic design. Therefore, as a part of a R&D-program supported by the government, impact tests on concrete beams, concrete slabs, and reinforcing steel as well as numerical investigations have been performed at BAM (1).

MODEL TESTS, A PART OF THE DESIGN CONCEPT

A reinforced concrete containment of a nuclear power plant cannot be tested to a scale of 1:1 under loading conditions like aircraft crash. However, merely the existing analytical methods are not sufficient to calculate the behaviour of RC-structures under impact loads up to large plastic deformations. Therefore, the design concept should be verified on the basis of both analysis and model experiment. To perform the required calculations, suitable mechanical models have to be available. Mechanical models for the calculation of concrete shells, slabs or beams are known, but they are normally based on experiments with static load conditions. For impact loads there are several uncertainties in modelling, for example the question whether the failure mechanism of an impacted RC-structural member is comparable to that under static loads, or the influence of high strain rates on the material behaviour. The investigation of these problems is one of the aims of the tests performed at BAM. If tests to a scale of 1:1 were not possible, tests in a practicable size would have to be carried out. They have to model the major phenomena of the structural behaviour in reality. It is supposed that RC-structural members which are loaded by a "soft missile impact" like aircraft crash will fail in punching shear, reaction shear or flexure with deformations up to and within the plastic range. Other local damage like scabbing, penetration or perforation are effects which belong to a "hard missile impact" and are not considered here. As known from static tests, the plastic flexural deformation will concentrate in the yield lines or plastic hinges. The rotational capacity of yield lines and plastic hinges as a function of the deformation rate and other parameters has to be examined for a wide range of applications. Therefore, we chose four-point supported RC-slabs as a model for RC-structural members with yield lines and simply supported RC-beams as a model for RC-structural members with plastic hinges. The influence of many parameters can be investigated by beam tests whereby a beam structure may be considered as a cut-out of a slab perpendicular to the yield line so that the condition at the plastic hinge may be transmitted to those in yield lines. The idea of our test concept is to reduce real structural behaviour to the main phenomena and to lay out the test as simple and clear as possible. Due to the fact that the deformation rate may influence the structural behaviour, it is necessary to have the same rate of straining in the composite reinforced concrete material during a test as in reality. This will occur if the time up to failure in a test and in reality has the same order of magnitude. This demand can be very well realized by the servo-hydraulic testing equipment at BAM, which was developed for these specific tests Fig. 1.

We decided to use original concrete and original reinforcing steel for all our tests to avoid uncertainties in the transmission to real structures. The material behaviour of reinforced concrete structures under impact and impulsive loads is unknown in a wide range of application.

Figure 1   Servohydraulic testing equipment at BAM for static and impact tests on RC-beams, RC-slabs and reinforcing steel

Thus, the dependency of many parameters on the strain rate cannot be modeled using micro-conrete or special reinforcement before having tested the original material behaviour sufficiently.

IMPACT TESTS ON RC-BEAMS, RC-SLABS, AND REINFORCING STEEL
Tests on RC-beams:
Up to now 56 simply supported RC-beams loaded by a single load in midspan have been tested at BAM (2), (3), (4). The span width of the beams varied from 3.20 m to 5.76 m and the measure of the cross section varied from 22 cm to 90 cm in height (25cm and 30 cm depth). All beams had sufficient stirrups to prevent

Figure 2   Centrally loaded RC-beam;   Figure 3   Measurement arrangements for impact tests on RC-beams
            flexural failure in a
            plastic hinge

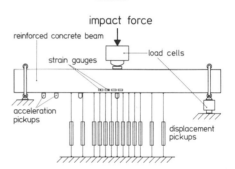

Figure 4   Impact force-time curve   Figure 5   Beam model for evaluation of the characteristic curve
            and midspan displacement-
time curve of a RC-beam impact test

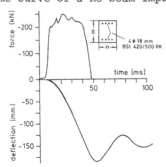
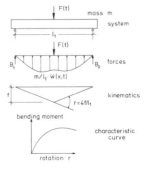

shear failure so that flexural failure occured in a plastic
hinge, Fig. 2. Measurement arrangements are shown in Fig. 3 and
Fig. 4 shows the impact force-time curve and the midspan deflection-time curve from one impact test. During the deformation
controlled tests, the measured data are stored in a computer.
Nearly all beams were deformed up to total failure. To study
the influence of the strain rate on the structural behaviour of
the beam, one static test and one impact test each was performed
per parameter value of the specimen. The investigated parameter
were the percentage of the upper and lower reinforcement, the
concrete strength, the type of steel reinforcement, the height
and span of the specimen, and the diamter of the reinforcement
bars. Comparing the deformations up to failure in static and
impact tests, an increase of the ultimate deformation in the
impact test can be stated. To receive the relation between
suitable load values and deformation values from tests, we have
to assume a model for the evaluation. Choosing the beam model
illustrated in Fig. 5, which contains the influence of inertial forces, the difference of the material behaviour in static and impact tests can be seen by comparing the corresponding moment-rotation relationship, Fig. 6. The moment-rotation
relationship, Fig. 6, shows the increase of the maximum bending moment and rotational angle in the impact test in comparison to the static test but no significant difference in the
crack pattern and failure mechanism could be stated comparing
both tests. Beams with a high percentage of the upper and lower
reinforcement failed by rupture of the reinforcement bars in
the tensile zone. In the event of only small upper reinforcement, failure of the compression zone occurs. The concentration of cracks in the plastic hinge, Fig. 2, implies a strain
concentration in the reinforcement. In Fig. 7, the permanent
strain distribution in the beam after a static and after an
impact test is plotted. A significant increase of the inelastic strains in the impact test can be observed. The tests
showed that the ductility of RC-beams is mainly influenced by
the mechanical properties of the reinforcing steel. Therefore,
additional tensile tests on the influence of the strain rate

Figure 6 Bending moment-rotational angle curves in static and impact tests

Figure 7 Permanent strains of the tensile reinforcement of similar RC-beams

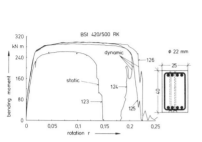
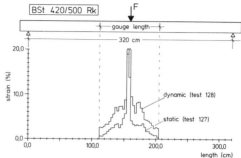

on the mechanical properties of reinforcing steel were initiated at BAM.

Tests on RC-slabs:

Static and impact tests on 35 slabs have been carried out at BAM. The size of the slabs was 3.00 m x 3.00 m, 16 cm and 22 cm thick. They were centrally loaded and supported at four corners on steel balls. In some tests the slabs were clamped in very stiff RC-beams at the edges. Fig. 8 shows the back face of a slab after an impact test. The yield lines can be seen very

Figure 8 Back face of a RC-slab with edge beams after impact test

Figure 9 Load-deflection curves of RC-slabs under static a) and impact b) load.

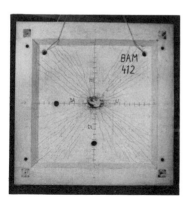
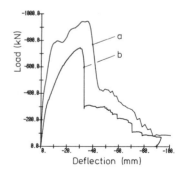

clearly. Three typical failure modes could be observed in the tests: "Primary punching failure", with only small bending deformations in the elastic range (real punching shear failure), "secondary punching failure", with greater plastic bending deformations and local bending cracks which induce the shear failure, and "overall flexural failure", whereby failure occurs because the ultimate rotational capacity of the yield lines is exceeded. In all cases significant strain rate effects could be observed, Fig. 9. In the tests the influence of several structural parameters such as amount of shear reinforcement, bending reinforcement, concrete strength, loading area, and boundary conditions have been studied. The evaluation of the slab tests is not yet finished. No special model exists for evaluation up to now, but the load-deflection-curves, Fig. 9, demonstrate the failure mode and the load bearing and deformation behaviour of the slab which is dependent on different test parameters.

Tensile tests on reinforcing steel

Tensile tests on three reinforcing steel types with different strain rates have been performed in a servohydraulic equipment, Fig. 1, (3). Many tests have been performed up to now to evaluate the dynamic properties of mild structural steel, but only

Figure 10 Influence of strain rate on stress-strain curves of reinforcing steel a) BSt 420/500 RK (cold worked) and BSt 1100 (high strength steel)

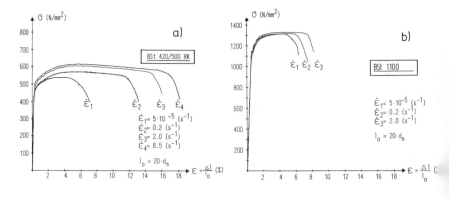

a few results were known from tests on reinforcing steel. In most of the former tests machined specimens were used whereas our investigations show that tests have to be performed without turning off the outer section of the bar. Some years ago it was almost a dogma that the characteristic values for the steel in the plastic range up to failure would decrease with increasing strain rates. This behaviour might be valid for structural steel or very high strain rates but for the reinforcing steel types investigated here a reverse behaviour has been found. Figure 10 a and 10 b show the stress-strain curves at different strain rates for a ribbed cold worked reinforcing steel (BSt 420/500 RK) and a ribbed high strength reinforcing steel (BSt 1100). It can be stated that the percentage permanent elongation increases with the strain rate in both cases but this effect is much more significant as far as the steel type BSt 420/500 RK is concerned.

FORMULATION OF RELIABLE DESIGN MODELS

Between test results and design requirements should be a close interaction. Design requirements are often formulated by assuming a specific model idea. For example, the rotational angle is limited for the "yield line" or "plastic hinges model" or the strain or curvature is limited in a plane cross-section or the shear stress is limited in a shear cone. In many cases it is possible to analyse tests on beams and slabs assuming different models if the required data could be measured. It might happen that some tests show phenomena that could not be modeled in the usual design model. New models have to be evaluated for this task or known models have to be modified. The reliability of the new model can be confirmed by comparison of test results and the results of computational investigations. For that reason, a numerical analysis of some RC-beams has been performed at BAM (5), (6). The Finite-Element-Code ADINA was used. Fig. 11 shows the FE-model used in the nonlinear dynamic analysis. To take into account strain rate effects, the concrete strength and steel yield stress was increased in comparison with static values. The increase of the steel yield stress was according to the results form our tension tests. This procedure is only

Figure 11 Finite Element Model of one half of the symmetric RC-beam

Figure 12 Midspan displacement-time curve for a RC-beam

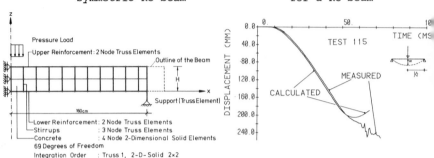

an engineering approach. As shown in Fig. 12 the computed midspan-deflection-time curve of that simple example is in good agreement with the experimental curve. For complex structures and load histories a great need for better formulation of material models considering strain rate effects or even strain rate history effects exists. The confirmation of those models requires further difficult experimental investigations. The material model used in the FE-model is based on the input of a stress-strain relationship. Other models are possible using a moment-curvature relationship or a moment-rotational-angle relationship in the evaluation. It is obvious that the model used for analysis of the tests has to correspond to the model used in design. Another aspect is that in design often different models are needed to represent reality, for example, one for the flexural behaviour and another for punching shear mode. The idea on how to formulate reliable design models often results from a precise analysis of tests.

CONCLUDING REMARKS

The results of the tests on RC-beams, RC-slabs, and reinforcing steel show that a reliable design of impact loaded reinforced concrete structures is only possible taking into account the observed phenomena. This means that tests are inevitable to observe the principle mechanism and to get reliable material data. If tests on natural size members are not possible, the reality has to be modeled in suitable tests on smaller specimens. For the evaluation of the tests advanced mechanical models are requested which can describe the most of the observed effects.

In case of impact loading of RC-structures design models based on static considerations are often insufficient. Furthermore strain rate effects or even strain rate history effects have to be included. A full consideration of all observed significant effects is not possible up to now. Therefore, more or less engineering approaches are requested which have to be verified in tests. Finally, it can be stated that there is a close interaction between tests and design models and both should be in good agreement with reality. The tests performed at the BAM are suitable for the determination of design parameters and the development of design concepts. Some results have already influenced German design codes, further evaluation is intended.

REFERENCES

(1)   BRANDES, K, LIMBERGER, E, HERTER, J, 'Kinetic Load Bearing Capacity of Impulsively Loaded Reinforced Concrete Members - Conception for Tests under closed-loop Control-Testing Facilities - Preliminary Tests on Reinforced Concrete Beams' (in German), BAM-Forschungsbericht 90, Bundesanstalt für Materialprüfung (BAM), Berlin, (April 1983).

(2)   LIMBERGER, E, BRANDES, K, HERTER, J, Berner, K, 'Kinetic Load Bearing Capacity of Impulsively Loaded Reinforced Concrete Members - Tests on Reinforced Concrete Beams' (in German), BAM-Forschungsbericht, Bundesanstalt für Materialprüfung (BAM), Berlin (in press).

(3)   LIMBERGER, E, BRANDES, K, HERTER, J, 'Influence of Mechanical Properties of Reinforcing Steel on the Ductility of Reinforced Concrete Beams with Respect to High Strain Rates', Proceedings of the RILEM/CEB/IABSE/IASS-Interassociation Symposium on Concrete Structures under Impact and Impulsive Loading, BAM, Berlin June 2-4,1982.

(4)   BRANDES, K, LIMBERGER, E, HERTER, J, 'Strain Rate Dependent Energy Absorption Capacity of Reinforced Concrete Members Under Aircraft Impact', <u>Transactions of the 7th SMiRT-Conference,</u> Chicago, 1983, Vol. J.

(5)   HERTER, J, 'Rechnerische Analyse des Trag- und Verformungsverhaltens von Stahlbetonbalken unter Stoßbelastung mit dem Programmsystem ADINA', <u>Berichte zum 12. Forschungskolloquium des DAfStb,</u> BAM, Berlin, 1981.

(6)   EMRICH, F, HERTER, J, PUFFER, G, 'Nonlinear Finite Element Analysis of Reinforced Concrete Beams under Impact Load in Comparison with Experimental Results', Proceedings of the <u>RILEM/CEB/IABSE/IASS-Interassociation Symposium on Concrete Structures under Impact and Impulsive Loading,</u> BAM, Berlin June 2-4, 1982.

DISCUSSION

Peter Barr   Atomic Energy Establishment   UK

Work on the impact resistance of concrete slabs faced with mild steel plates has been carried out by the UKAEA at Winfrith, and was reported in paper J8/4 of the 7th SMIRT conference held in Chicago in August 1983.

Section 2   Papers 9–16

Commentary by Dr G. SOMERVILLE
*(Director of Research and Technical Services,
Cement and Concrete Association, UK)*

In summing up the papers in this Section I would like to make some general points in an attempt to give a sense of perspective to the subject matter of this section, namely, the reliability and accuracy of models.

All design, and most research, is about modelling in some form. This can involve mathematical or analytical modelling. It can involve the derivation of empirical design equations based substantially on test results from the laboratory (with no mention made of "size" or "scale") or it can involve physical testing to a reduced scale (the subject of this seminar). All of these approaches have inbuilt inaccuracies of one sort or another. The choice of which technique to use in a particular case depends primarily on the objective of the task in hand. Requirements for accuracy will vary considerably being quite different if models are being used as part of a general research programme compared, say, with simply establishing whether or not performance is better than a stated minimum requirement. This point is central to this seminar and, indeed, to the whole art of modelling.

My next point relates to the references given at the end of each of the 30 odd papers presented, which can be a very useful indicator of how developments are taking place. I note again the normal tendency of all authors to be somewhat insular in that they tend to refer only to papers published in their own country and in their own language. It is important that events of this type are held fairly frequently to help break down these barriers. There is also a tendency not to go far enough back in literature. Today, for example, I see very few references prior to 1970. This means that we tend to reinvent the wheel every 15 to 20 years. In preparation for this seminar, I reread the papers presented in ACI's Special Publication No 24 which was published in 1970. In so doing, I find that we are repeating some of the earlier mistakes, but we are also adding to knowledge. In today's papers, for example, we have now fully appreciated the significance of strain gradients in influencing model results and we now know a great deal more about the effects of fabrication and construction techniques. I also noted that some original modelling techniques have been developed. The other point that clearly emerges from comparing sets of papers written at an interval of 15 years is that the priorities are different. Previously we gave equal emphasis to elastic models and so called 'realistic' models. Elastic models have now virtually disappeared having been substantially overtaken by developments in computer techniques. Moreover, the usage of models has changed. It is much more common these days to use models as part of general research and development work, and that change in use should be firmly borne in mind in deciding what accuracy and reliability we want.

My third point is that if we are using models as part of general development work in design, then the translation of the results obtained into practice becomes very important. In this context we must remember that we are trying to model structures built in concrete which is a variable material and can result in differences in performance in practice. Some of these differences are inherent while some are due to factors such as workmanship. Variability is therefore an issue in translating the data so obtained and helps put physical modelling into perspective. Indeed, it is one of the key issues in deciding between mathematical and physical models and hence on the accuracy required for design purposes.

Finally, I am very pleased indeed to see that the interest in modelling is as strong as ever. All the problems are still not solved. Bond remains a continuing problem not only in influencing deformations and cracking but also with the modelling of dynamic effects, where the strain rates are critical in affecting the reinforcement and hence the results obtained. The awareness that derives from working with very small scale specimens is crucial in using the results subsequently for general design purposes. Because of that it is essential that we not only continue to work with small scale specimens but we also continue to develop the techniques themselves.

17 Modeling punching and torsional shear at penetrations in R/C slabs

WOO KIM and RICHARD N. WHITE
Cornell University, Ithaca, New York, USA

ABSTRACT

Small-scale models are used to study the behavior of steel penetrations through reinforced concrete slabs, with the penetrations loaded in pure torsion and in combined torsion and punching action. Techniques are given for modeling shear studs, for making quick-curing model concrete, and for loading penetrations under combined torsion and punching shear. The results of 22 tests on 1 in. thick slabs are summarized.

INTRODUCTION

Punching action on penetrations in reinforced and prestressed concrete pressure vessels is a common design loading. Methods for dealing with these loads are reasonably well-defined except for cases where there are large in-plane tensile forces existing in combination with punching shear, as may occur in reinforced concrete nuclear containments. Torsional loads acting tangent to the wall of the vessel, on a steel penetration through the wall thickness, also occur in nuclear containments. Design approaches for torsion are based strictly on elastic stress calculations with no confirming experimental evidence to justify this approach. Combined punching and torsion loads on penetrations represent an area of unknown behavior, with the degree of interaction between the two loadings not understood.

This research program is directed at studying the behavior and strength of small-scale concrete slabs with penetrations loaded in torsion alone, in punching shear alone, and in certain combinations of torsion and punching shear. Steel penetrations with two different d/t ratios (0.5 and 1.0) are anchored to the 1 inch thick model concrete slab with small-scale shear studs. Other variables include concrete strength, degree of in-plane concrete confinement, and ratio of torsional shear to punching shear.

Modeling techniques for very small-scale shear studs are described, including two methods for attaching the 1/4 inch long studs to the metal penetrations, the use of a gypsum model concrete to permit testing within 24 hours after casting, and features of the special loading system for applying simultaneous punching shear and torsional shear.

Detailed results are presented in References (1) and (2).

TEST SPECIMENS AND EQUIPMENT
Geometry: The test specimens modeled a thick slab with a steel pipe penetration. Hence, they had relatively small ratio of span to depth and large reinforcing steel ratios. In particular, the specimens displayed several similarities to the wall of the nuclear containment vessel but they were not intended to be a scaled version of the wall.

Eight of the 10 specimens loaded in pure torsion were 8 in. square and 1 in. thick, with either a 1 in. diameter or 0.5 in. diameter steel penetration sleeve which was anchored to concrete by stud shear connectors as shown in Fig. 1. Model reinforcing steel, 0.11 in. diameter, was used in both directions and in both faces of the slabs. The reinforcement ratios, calculated in the usual manner for a slab were 0.0222 and 0.0194 in the two directions. The average effective depth, d, was 0.79 in. The 1.0 diameter penetration sleeve had 64 studs which were 0.048 in. diameter and projected 1/4 in. from the surface of the penetration. The 0.50 in. diameter penetration had 26 studs of the same dimensions.

Two torsional shearing specimens had reduced width in one direction and greatly reduced reinforcement ratios to study the influence of in-plane stiffness on torsional shear; they are described in Refs. (1) and (2).

The 12 specimens for the combined torsional and punching shear study were the same as described above except a flange was added to the penetration on the punching load side of the slab. The flange had an outside radius 0.25 in. larger than the penetration radius and acted as a loading pad when punching load was applied. The specimen geometries are summarized in Fig. 2.

Model Concrete: Ultracal 60* gypsum model concrete was chosen for use in this modeling study because of its several advantages. Design compressive strength is achieved in less than one day; hence tests can be performed 24 hours after a specimen is cast. Mixes of suitable strength have a relatively high ratio of gypsum concrete to sand, which gives a highly workable mix. Since the level and degree of cracking was of primary importance in this study, the model concrete tensile strength had to simulate the tensile strength of prototype concrete as closely as possible. While model concretes tend to have tensile strengths that are too high, gypsum model concrete is better than Portland cement model concrete in this respect. On the other hand, excessive and differential drying of gypsum model concrete, if not controlled, can produce brittle and unrealistic behavior. But this was not a significant problem in this study because the slab models were of uniform thickness and uncomplicated in form.

The model concrete was made with a natural mortar sand. Proportions and strengths for four mixes were:

Ultracal 60*	Sand	Water	f_c', psi
1	4	0.5	1700
1	4	0.425	2640
1	2	0.35	3770
1	1.25	0.325	4400

A typical stress-strain curve for gypsum model concrete, measured on a 1 in. by 2 in. cylinder, is given in Fig. 3. Average tensile splitting strength was 6.68 $\sqrt{f_c'}$. Additional information on gypsum model concretes is given in Ref. (3).

---

* A product of the U.S. Gypsum Company

Model Reinforcement: The reinforcement steel used in all slabs was commercially-available small bars with cold-rolled surface deformations. These bars were 0.11 in. diameter and were annealed to a yield strength of 32 ksi (see Fig. 4).

Model Shear Studs: Commercially available nails (so-called finishing nails) were used for model studs. These were 0.048 in. diameter and 1.0 in. long. The ultimate tensile strength and shear strength were determined by testing a number of nails. The resulting average ultimate tensile stress was 110 ksi, and the ultimate direct shear strength was 150 lbs per nail.

In comparison to the shear connectors used on penetrations in nuclear reactor vessels, both model stud diameter and stud length were distorted from true similitude. Furthermore, the percentage of stud area with respect to the interface area was 4% in the models, and about 0.2% in the typical prototypes. The higher percentage in the models was chosen to enable higher loads to be transferred into the concrete in an attempt to make the concrete critical for controlling strength.

Studs were connected to the penetration with techniques developed specially for this study. For the 1 in. diameter hollow penetrations, 0.048 in. diameter holes for each stud were drilled at the exact points through the penetration wall, and then a shortened nail was driven into the hole with a hammer. Next, the protruding ends of the nails inside the sleeve were bent, and the inner hole of sleeve was filled with a steel-to-steel epoxy, as shown in Fig. 5. In the 0.5 in. penetrations, the full length nail was driven completely through the two walls, with holes located in a staggered fashion to avoid interference of the nails. These techniques enabled full shear strength development of the model studs.

Loading Device: A double reaction frame was designed and built to apply the torsional and punching force to the specimens simultaneously. The punching load was applied by a small hydraulic jack and the torsional load was applied by a mechanical loading device positioned in a horizontal orientation which pushed the torsion arm welded to the mid-height of the vertical ram, as shown in Fig. 6. A steel pad, a half in. diameter ball, a frictionless ball bearing, a bearing house, and the punching load cell were inserted between the jack and the vertical ram. The ball was used to eliminate any eccentric loading due to unexpected misalignment. The ball bearing played the role of eliminating the frictional force which might result when punching loading and torsional loading were applied at the same time. A relatively thick bearing house between the ball bearing and the punching load cell was used to ensure that uniform stress was transferred to the load cell which was positioned beneath the bearing house. The punching load cell was designed to have high sensitivity and was made of high strength steel. A special connection system which looked like a universal joint was made at the lower end of the vertical ram. The punching load was transmitted through a 0.5 in. diameter ball which was inserted between bottom end of the ram and top end of the penetration sleeve. This ball also eliminated any eccentric vertical load. The torsional load was transmitted through the engagement of the pin, passing through the penetration sleeve diagonally, and the teeth typed cylinder, welded to the vertical ram. A test in progress in shown in Fig. 7.

RESULTS AND CONCLUSIONS

Only the main conclusions about specimen behavior will be given here; details are presented in Refs. (1) and (2).

1. Torsion alone: For penetrations with no external sleeve, failure was by concrete splitting along the outermost rows of shear studs, loss of cover, and reduction in load capacity, as shown in Fig. 8. There was significant ductility after the appearance of the first crack. Ultimate shear stresses calculated at the face of the penetration ranged from 0.35 f'_c to 0.69 f'_c (23 $\sqrt{f'_c}$ to 35 $\sqrt{f'_c}$, with f'_c in psi), with the higher values for the 1/2 in. diameter penetration. Shear stresses on the surface defined by the outer ends of the studs ranged from 7.2 to 15.2 $\sqrt{f'_c}$. There was some shear failures of studs on the inner rows in specimens with higher strength concretes. There was very little evidence of distress in the concrete except for the major cracks induced by the splitting forces from the loads on the outer rows of studs. The addition of a flange on one side of the slab prevented a complete splitting failure on that side and increased torsional shearing strength by about 20%.

2. Combined torsion and punching: Because the mode of failure from pure

punching action has little relation to the torsional failure mode, there is relatively little interaction between the two types of loading, and the interaction curve is nearly square. Plotting P/P_o vs. T/T_o, where T represents torsion and P represents punching, a peak value of P/P_o of about 0.8 can be reached with T/T_o equal to 1.0. The loading path followed in proceeding to failure under combined punching and torsion seems to have little effect on results. Strength in pure punching was somewhat higher (9 to 10 $\sqrt{f_c'}$) than one would expect in prototype slabs.

3. The small scale modeling techniques introduced for this study gave consistent and repeatable results, and it is believed that the behavior shown in these 22 experiments provides substantial new insight into how a prototype structure would behave. Potential size effects have not been investigated, but it is expected that they might be substantial for punching strength values, as shown by other investigators.

REFERENCES

(1) KIM W., Combined Punching Shear and Torsional Shear in Reinforced Concrete Slabs, M.S. Thesis, Dept. of Structural Engrg., Cornell Univ., Ithaca, N.Y., (Aug. 1984).

(2) WHITE, R. N. and KIM W., "Structural Behavior of Penetrations in Reinforced Concrete Secondary Containment Vessels", Proc. of Second Workshop on Containment Integrity, U.S. Nuclear Regulatory Commission, (to be published as NUREG Report in 1984).

(3) SABNIS, G. M., HARRIS, H. G., WHITE, R. N., and MIRZA, M. S., Structural Modeling and Experimental Techniques, Prentice-Hall, Inc., Englewood Cliffs, N.J. (1983).

ACKNOWLEDGEMENTS

This research was conducted in the Structural Models Laboratory at Cornell University. The senior author was supported by the Education Ministry of Government of Korea.

Fig. 1 - Specimen Geometry

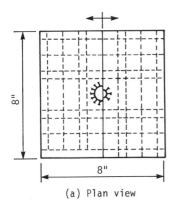

(a) Plan view

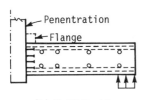

(b) Half Section

Fig. 2 - Test Specimens

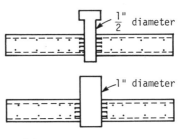

(a) Torsional test series

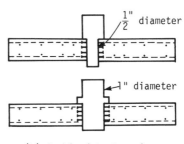

(b) Combined test series

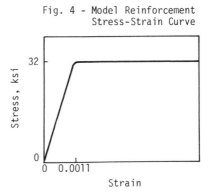

Fig. 3 - Model Concrete Compressive Behavior

Fig. 4 - Model Reinforcement Stress-Strain Curve

Fig. 5 - Model Shear Stud Details

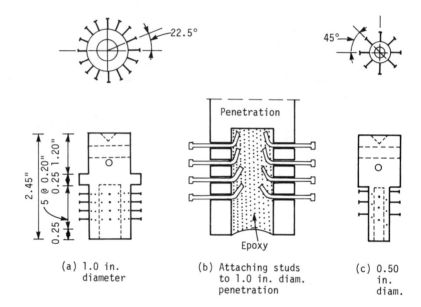

(a) 1.0 in. diameter

(b) Attaching studs to 1.0 in. diam. penetration

(c) 0.50 in. diam.

Fig. 6 - Loading Device

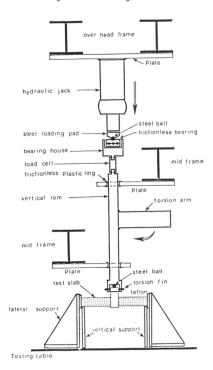

Fig. 8 - Typical Behavior in Torsion

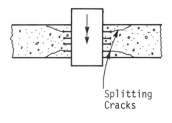

Splitting Cracks

Fig. 7 - Test in Progress

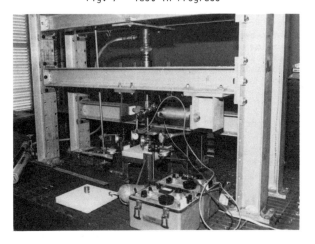

DISCUSSION

Dr F K Garas   Taylor Woodrow Construction Ltd   UK

From your work can you comment on the effects of varying the thickness of
the penetration liner on the shear strength. Can you quantify the
contribution of the penetration liner towards the ultimate strength?

Author's Reply

It is not clear as to what is meant by "penetration liner"; perhaps the liner
referred to here is the flange that bore directly on the concrete slab on its
exterior surface. The thickness of the flange was not varied, and tests were
not done without the flange in place (only torsional shearing tests were
done with the flangeless penetration). The general effect of the flange is
to shift the punching shear failure surface out away from the penetration.
In effect, the outer diameter of the flange defines the loaded area in just
the same way that a column bearing on a slab would define a critical area for
punching shear. The flange diameter controls punching strength.

If the thickness of the "penetration liner" refers to the thickness of the
penetration itself, the geometries used would not reveal any dependence on
shear strength since the smaller penetration was a solid 1/2 inch diameter
steel rod, and the larger penetration was a very thick-walled 1 inch diameter
steel cylinder. No attempt was made to model the relatively thin and flexible
penetration of the type used in prototype nuclear containment vessels. Thus
the question of potential tearing of the penetration material itself prior to
failure of the shear studs, or the surrounding concrete, was not addressed in
this model study. From the behaviour observed, it is believed that failure
in the concrete around the penetration itself (in the locality of the studs)
would never occur, and that punching strength would be governed by overall
punching effects (function of the flange diameter), shearing of the studs
(for flangeless penetrations with relatively small amount of stud area), or
by possibly tearing of the penetration material away from the studs. The
precise mode of load transfer into the penetration/concrete interface would
have to be specified in trying to predict prototype punching shear strength
at a penetration.

Dr S H Perry   Imperial College   London   UK

Does repeated cycling of either the axial load or applied torque affect the
general independence of the punching shear strength of the applied torque?

Author's Reply

While we have not done any cyclic loading of the test specimens, the different
failure modes observed for the two types of loading lead us to believe that

the interaction between the two modes of behaviour will be weak for cyclic as well as for single loads. The additional degree of cracking and degradation of concrete around the shear studs should produce slightly weaker (and certainly more flexible) behaviour in torsion when loads are cycled many times, but this trend in response should have little, if any, influence on the capacity in punching shear strength.

The question is an excellent one and perhaps we can perform some additional tests to have a better answer than that speculated above.

18 Physical and analytical models for flat slab/edge column connections

D. J. CLELAND, S. G. GILBERT and A. E. LONG
Department of Civil Engineering, Queen's University, Belfast, U

SUMMARY
Elastic plate bending finite element analyses have been used to model the behaviour of edge panels of a flat slab structure. Modifications, to allow for cracking, have been made based on observations of experimental behaviour and engineering judgement. The method has been found to show good correlation with results from physical models with carefully controlled boundary conditions. A simplified equivalent frame method has been derived using the mathematical mod

INTRODUCTION
The stress condition which exists in a slab at the junction with a column is extremely complex. Researchers have used intuitive basis to formulate methods for estimating the shear stresses and bending moments around the column. Fini elements offer an alternative for determining the stresses but clearly elastic plate bending finite element analyses are only appropriate at low levels of applied load.

The current investigation was concerned with two objectives:
1. To determine the stresses in the vicinity of an edge column at failure so that a criterion for failure could be established.
2. To establish a method of predicting the distribution of moments within a slab panel at all stages of loading up to failure.

To satisfy these objectives it was decided to approach the investigation on two fronts. Firstly, an experimental investigation was carried out to determine accurately the behaviour of slabs under vertical load with particular reference to location and stage of cracking. The second part was an analytical study in which an attempt was made to simulate the observed behaviour of the experimental structure.

EXPERIMENTAL PROGRAM
Test Models
Carefully scaled full panel test models incorporating two edge columns were used. The details of the model and the basis for the selection are illustrated in Fig. 1. The achievement of correct boundary conditions is of paramount importance. This accounts for the incorporation of two columns which ensure continuity in the direction perpendicular to the free edge similar to that which exists in a real structure. In the orthogonal direction (i.e., parallel to the free edge) continuity was simulated by using a system of edge restraints similar to those used by Long and Masterson (1).

The test models were approximately one-third scale with concrete aggregate and reinforcement sizes scaled accordingly. Reinforced concrete and unbonded post-tensioned concrete models were tested under static load conditions. At each stage of the loading sequence measurements were made of slab deflection, column moments, reinforcement stresses and slab rotations.

Description of Behaviour
Cracks first appeared in the slab at the inner corners of the columns. These hairline cracks occurred at about design service load or slightly higher in the case of the post-tensioned slabs. The application of further load caused these cracks to meet across the inner face and shortly after torsional cracks developed; extending from the initial cracks to the free edge at $45°$ (Fig. 2).

Radial cracks from the columns, in some cases meeting those from the opposite column, also developed. In the reinforced concrete slabs this took place at about the same stage as the torsional cracks while in the post-tensioned slabs it was delayed until near failure. Similarly cracks also occurred on the underside of the slabs at mid-span locations in both directions although not until just prior to failure in the post-tensioned slabs. In all cases there was the occurrence of sudden punching shear failure at one of the columns.

Column Moments

The moments transferred to the columns were determined by measuring the horizontal reaction at the extreme pinned ends of both the upper and lower columns. Interpretation of these horizontal reactions required care because lateral expansion of the slab affected the distribution of moment between the upper and lower column particularly near failure. Typical experimental relationships between total moment and axial load are shown in Fig. 3.

Generally the change in slope at fairly low loads corresponds to the occurrence of cracking in the column region while the later increase in slope tends to coincide with mid-span cracking. The influence of the latter is not so significant in reinforced concrete slabs possibly due to the more controlled crack development resulting from higher percentages of bonded reinforcement.

ANALYTICAL STUDY

Finite Element Program

A plate bending finite element analysis program employing 12 degree of freedom rectangular elements was used. An element grid of the type shown in Fig. 4 was considered suitable for one-quarter of the model structure since the area in the vicinity of the column was of primary interest. Two methods were considered for representing the finite sized column:

1. The column can be simulated by an arbitrary thickening of the slab elements immediately over the columns with the column stiffness (axial and bending) concentrated at the centre node.
2. Point supports (axial stiffness only) can be located at the corners of the physical column as in the method outlined by Joffriet (2). The axial stiffness of these supports can be chosen so that the overall moment-rotation relationship is satisfied.

The second approach was preferred because the discontinuity in stiffness in the first method resulted in stresses at the column boundary which are difficult to interpret. Furthermore the latter approach was assumed to pertain to the situation after initial cracking had occurred at the corners of the columns.

Using the column moments as the criterion it is clear that the finite element analyses do not agree with the experimental results except at very early stages of loading (Fig. 3). A better agreement with experimental observations was then sought by altering the stiffness of specific elements to represent the effect cracking.

Flexural Cracking near the Column

The first method of modelling this crack which was considered was to reduce the stiffness of elements along the inner face of the column location. This direct method proved to be unsatisfactory because again the discontinuity affected the moments and shears in the region of primary interest. It was decided to achieve a similar effect by reducing the stiffness of elements at the centre of the column location (Fig. 4). The amount by which the stiffness was altered was based on a beam analogy and the length over which the crack was assumed to affect the stiffness was estimated, after consideration of the literature (3), at 16 times the bar diameter. (The flexural rigidity was assumed to remain constant between cracking and yielding based on a simple three stage linear moment-curvature relationship (Fig. 5)).

Torsional Cracking

During the tests torsional cracking was observed to occur quite soon after the initial flexural cracking. The effect was to reduce the degree of rigidity of the connection to that approximately comparable to a slab connected only along the inner column face. The finite element representation adopted was to assign a value of zero to the torsional stiffness of the elements adjacent to the side faces of the column. The value of zero was based on the fact that the section contains little or no torsional reinforcement and is consistent with the assumption made by other researchers (4). The flexural rigidity of these elements was also reduced to a value consistent with their cracked stiffness.

Mid-span Cracking

This was accounted for by reducing the flexural rigidity of the appropriate elements in one direction (Fig. 4).

Crack Loads

Having established the form of cracking and the stiffnesses relevant to the post-cracking stage it remained to determine an indication of the load or stress levels at which the various forms of cracking take place. It is reasonable to expect that flexural cracking may occur when the extreme fibre stress reaches a value equal to the modulus of rupture. However since flexural cracking and torsional cracking at the column are closely related a single criterion was established i.e. an extreme fibre tensile stress at the inner column face of $1.35\sqrt{f_{cu}}$ (50% above the modulus of rupture). For cracking at mid-span an extreme fibre tensile stress equal to the modulus of rupture, i.e. $0.9\sqrt{f_{cu}}$

Comparison of Analytical Model with Test Results
The above method has been based primarily on post-tensioned slabs and the typical degree of correlation with test results as measured by the moment v. axial load relationship is shown in Fig. 6.

USE OF ANALYTICAL MODEL
Effective Stiffness Approach
To avoid having to superimpose various analyses for each stage the method was altered to use effective flexural and torsional stiffnesses (Fig. 5). Using these single linear relationships between moment and curvature (i.e. stiffness) single linear relationships between column moment and axial load can be obtained

Theoretically these relationships should be exactly correct at only one load so it was considered appropriate, with modern emphasis on the ultimate limit state, that this should be the yield or failure load. Thus the effective cracked flexural rigidity of the section was taken as the ratio of the yield moment to the curvature at first yield (Fig. 5). Similarly, based on experimental observations, of rotation, the effective torsional stiffness was found to be approximately one-quarter of the uncracked value.

Cracking at mid-span has not been taken into account since:
  (a) it occurs near failure and therefore has minimal influence
and (b) its influence can be included in the normal redistribution procedure.

Equivalent Frame Method
Using effective stiffnesses based on typical levels of reinforcement a parametr study was carried out. From this an expression for the equivalent column stiffness has been obtained for use in a modified frame method. The method is compared with the results of the model tests in Fig. 3. As expected the correlation is good for the post-tensioned slabs on which the method is based but not so good for the reinforced concrete slabs. This may be because cracki initiates at lower load levels and is more extensive in reinforced concrete sl

To improve the correlation the procedure could be revised to use lower values of effective stiffness. However it was decided that such a level of precision was not practical or warranted and in any case the single expression for the equivalent column stiffness gives results which are only marginally conservative for reinforced concrete slabs.

Elsewhere (5) the method is compared with the ACI (318-77) and CP110 (1972) frame methods and is found to agree closely with the former while predicting column moments up to 45% less than those predicted by CP110 (1972).

CONCLUSIONS

The following conclusions and observations can be drawn from this integrated experimental and analytical study.

1. Flexural and torsional cracking in the vicinity of the column were found to have a significant influence on the behaviour of the models.
2. A relatively simple analytical procedure which is calibrated in the light of experimental results provides a useful means of studying the complex slab-column connection.
3. As a consequence an indication of the critical stresses at the column face at failure has been achieved and an equivalent frame method has been devised.

REFERENCES

1. Long, A.E. and Masterson, D.M., "Improved experimental procedure for determining the punching strength of reinforced concrete flat slab structures", *Shear in Reinforced Concrete SP-42*, Vol. 2, ACI, Detroit, 1974.
2. Jofriet, J.C., *Analysis and Design of Concrete Flat Plates*, Ph.D. Thesis, The University of Waterloo, Ontario, October 1971.
3. Hsu, C.T. *Investigation of Bond in Reinforced Concrete Models*, M.Eng. Thesis, McGill University, Montreal, April 1969.
4. Kemp, E.L. and Wilhelm, W.J. "Influence of spandrel beam torsion on slab capacity based on yield line criterion", *Analysis of Structural Systems for Torsion, SP-35*, ACI, Detroit, 1973.
5. Long, A.E. and Cleland, D.J. "An equivalent frame method for slab column structures", *The Structural Engineer*, Vol. 59A, No. 5, (May 1981).

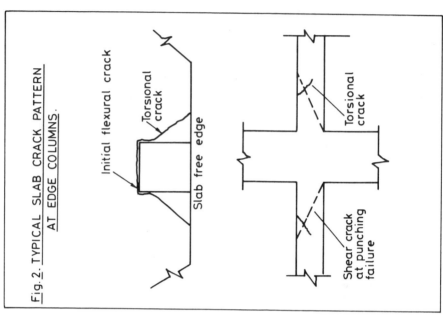

Fig. 2. TYPICAL SLAB CRACK PATTERN AT EDGE COLUMNS.

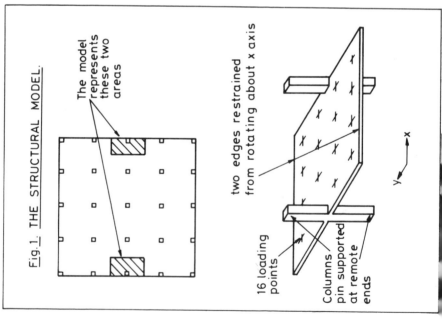

Fig. 1. THE STRUCTURAL MODEL.

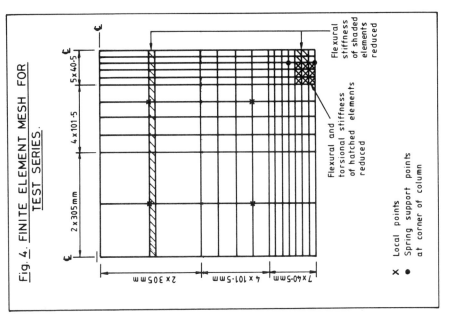

Fig. 4. FINITE ELEMENT MESH FOR TEST SERIES.

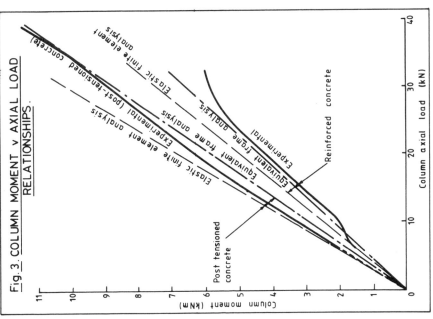

Fig. 3. COLUMN MOMENT v AXIAL LOAD RELATIONSHIPS.

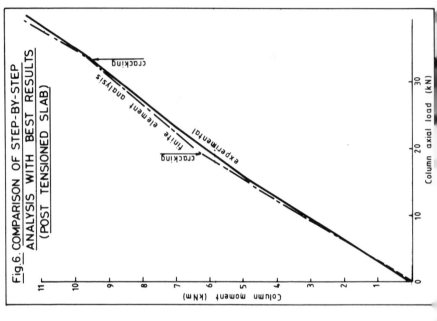

Fig.6. COMPARISON OF STEP-BY-STEP ANALYSIS WITH BEST RESULTS (POST TENSIONED SLAB)

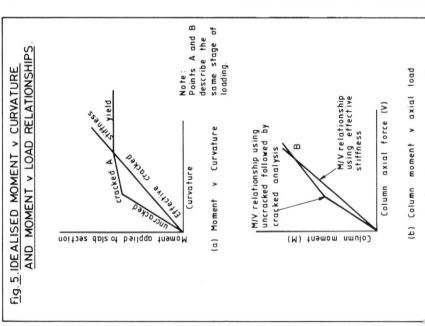

Fig.5. IDEALISED MOMENT v CURVATURE AND MOMENT v LOAD RELATIONSHIPS.

DISCUSSION

Dr R J Cope   University of Liverpool   UK

A thin plate finite element formulation has been used to analyse the models. This approach to plate analysis is known to give incorrect values for twisting moments and shear forces near a free edge. To obtain improved values, the twisting moments and shear forces predicted to act on the free edge can be replaced by statically-equivalent stress-resultants on sections intersecting the free edge[D1].

A comparison of the twisting moments and shear forces on such a section is shown in Fig D1. Details of the analyses used are given in reference D1. The results of the thick plate theory analyses satisfy the boundary conditions. They suggest that the unmodified thin plate theory predictions are grossly in error over a section between hand 1.5h wide.

In the paper, the results of thin plate theory are used directly. Also, the twisting stiffness coefficient is reduced in the edge zone to obtain a good fit between the analytical and theoretical results. Since the stiffness directly influences the fictitious twisting moments predicted by thin plate theory, it is difficult to realise the quantitative physical significance of the values used.

One way of improving the analytical procedure would be to use a thick plate finite element formulation. That would be expensive, however, and could perhaps, only be justified if a non-linear finite element procedure were used to trace the spread of cracking and material degradation.

A cheaper alternative would be to include beam elements along the edges of the slab elements. The torsional stiffness of the "edge strips" could then be modified directly. Following that line of thought, the most economic solution might be to use a grillage idealisation of the entire slab.

Figure D1   Twisting Moment and Shear Forces

on a Section Intersecting a Free Edge

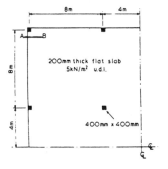

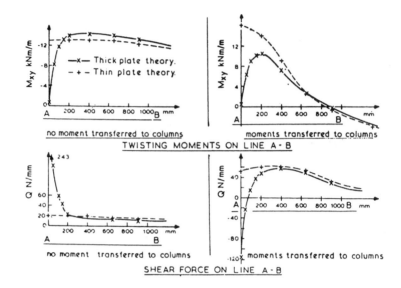

REFERENCE

(D1) COPE, R J and RAO, P V "Shear forces in edge zones of concrete slabs" The Structural Engineer, Vol 62A, No 3, 1984, pp 87-92.

Author's Reply

Dr Cope's comments and reference to his recent paper on the various methods of analysis, ie thin plate and thick plate finite element and grillage are interesting.

In response to his more specific points the authors would make the following observations:

(1) Typical distributions of twisting moment and shear along a section at the side of the column d/2 away is shown in Fig D2. This indicates that, in agreement with Dr Cope's observations, a considerable proportion of the column fixing moment is transferred to the slab by the out-of-balance distribution of shear forces. The error which exists at a free edge is clearly less significant at this critical section close to a column support.

(2) The occurrence of a torsional crack is considered to be a compatability requirement. (D2) It has been assumed to be associated with flexural cracking at the front face of the column so that the actual values of twisting moment in the edge strip have not been used directly in the analysis. Interpretation of these moments does not then arise.

(3) After cracking the torsional stiffness was assigned a value of zero which has a definite physical significance in spite of the fact that the twisting moments may be in error near the free edge.

(4) As stated in the paper the method of analysis was used to determine critical stresses at failure in order to establish a realistic failure criterion. The critical stresses were considered to be bending moments along the front inner face of the column. Thus shear stresses/forces although computed by the program were not used in the study.

Work has been carried out at Queen's University, Belfast, on the use of grillage analysis for flat slabs (D3) and this indeed confirms Dr Cope's contention that this is a more economic method of analysis.

D2   Collins, M P and Lampert, P "Redistribution of moments at cracking - the key to simpler torsion design", Analysis of Structural Systems for Torsion, SP-35, ACI, Detroit 1973.

D3   Rankin, G I B, A parametric investigation of flat slab behaviour by grillage analysis, Honours dissertation, Department of Civil Engineering, QUB, 1978.

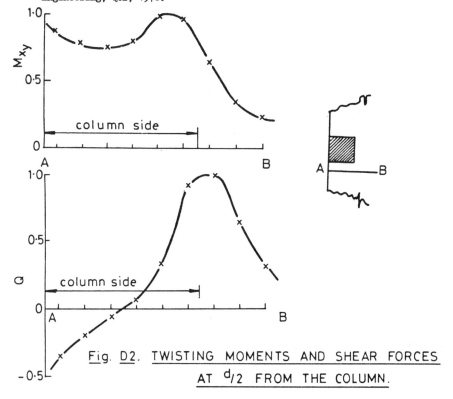

Fig. D2. TWISTING MOMENTS AND SHEAR FORCES AT $d/2$ FROM THE COLUMN.

DISCUSSION

Dr F K Garas   Taylor Woodrow Construction Ltd   UK

One of the problems in testing flat slab models is how to simulate boundary conditions as they occur in the elastic/plastic region - boundary conditions can influence the level at which cracking can occur and also the ultimate strength. Would you please comment on your solution to this problem?

Author's Reply

Dr Garas has focussed on an important aspect about model testing, not just of flat slabs but of any structure which is statically indeterminate or continuous. In these cases it is not sufficient just to ensure that the same state of equilibrium exists in the model as has been found from an elastic analysis of the prototype.

For example with the correct boundary conditions in-plane forces which can promote or delay cracking may arise. Compressive membrane effects due to the areas of slab around the column zone may not be obtained if the size of the model is less than a full panel.

Further unless the prototype boundary conditions are precisely modelled from the outset of the test the stresses after cracking and/or yielding will not be valid since the effects of redistribution will not be correctly simulated. In the case of a flat slab redistribution of moments from the highly stressed column region to mid-span has a significant effect on the ultimate strength.

These aspects greatly influenced the choice of boundary conditions for the physical models described in the paper.

1. Rankin, G I B, "Punching failure and compressive membrane action in reinforced concrete slabs", Ph D Thesis, QUB 1982.

DISCUSSION

Dr W Gene Corley   Prestressed Concrete Association   USA

The "Effective Stiffness Approach" described in the paper was developed for slabs with no edge beams and no interior beams. How would the simplified "Equivalent Frame Method" apply if beams were present?

Author's Reply

As Dr Corley has correctly stated the effective stiffness or equivalent frame approach outlined in the paper was developed for slabs with neither edge nor interior beams. However, the method has been applied to flat slabs with edge beams by substituting beam depth for slab depth in deriving the factor which modifies the stiffness. It has been found that provided the beams are not deep (ie depth column side) the predicted moments agree quite closely with those derived from the ACI 318-77 approach and with the limited experimental results available[2].

1. Long, A E & Cleland, D J "An equivalent frame method for slab-column structures", The Structural Engineer, Vol 59A, No 5, May 1981.

2. Long, A E, Cleland, D J & Kirk, D W "Moment Transfer and the Ultimate Capacity of slab column structures", The Structural Engineer, Vol 56A, No 4, April 1978.

Author's Closure

Table D1 shows the results of the ultimate load tests carried out at Queen's University on six reinforced concrete edge column tests, five post-tensioned edge column tests and three interior column models in post-tensioned concrete subject to transfer of moment. The results are compared with the predictions based on CP 110 (1972) [1] and The Concrete Society Report [2] and also the ACI design method [3].

Table D1 Comparison of Test Results with Code Predictions

Model Ref*	Test Punching Load P_T	Predicted capacity P_p (kN)		P_T/P_p	
		CP 110	ACI 1977	CP 110	ACI 1977
E1.R	66.2	37.8	43.8	1.75	1.51
E2.R	50.1	26.4	22.9	1.89	2.18
E3.R	32.7	22.8	17.7	1.43	1.84
E4.R	37.1	22.7	15.5	1.64	2.40
E5.R†	31.8	13.0	13.6	2.45	2.34
E6.R	37.6	22.9	16.8	1.64	2.24
E1.P	47.0	32.0	20.0	1.47	2.35
E2.P	49.0	32.0	20.0	1.53	2.45
E3.P	56.0	32.0	23.0	1.75	2.43
E4.P	40.0	32.0	17.0	1.25	2.35
E5.P	52.0	32.0	19.0	1.62	2.73
I1.P	100	42.0	56.0	2.38	1.78
I2.P	103	42.0	60.0	2.45	1.71
I3.P	75	26.0	47.0	2.88	1.60

*First letter refers to Edge or Interior column, second letter refers to Reinforced or Post-tensional model

†Column in Model E5.R was outside edge of slab

*First letter refers to Edge or Interior column, second letter refers to Reinforced or Post-tensional model

+ Column in Model E5.R was outside edge of slab

From Table D1 the following observations may be made:

(1) The ratio of test punching capacity to that predicted by the CP 110 method varies between 1.25 and 2.88. The corresponding range fro the ACI method is 1.51 to 2.73.

(2) The ratio of test to predicted punching capacity by the CP 110 method, is generally greater for interior columns subject to transfer of moment than for edge columns. This is primarily due to the magnification factors for the effects of moment transfer on the shear stress.

$1 + \dfrac{12.5M}{VL}$ for interior columns

1.25 for edge columns

The reverse is the case for the ACI method.

(3) Model E5.R in which the column was only connected to the slab along the inner face exhibited only minor reduction in capacity compared to the corresponding slab with torsional edge strips (Model E4.R). The CP 110

predicts a considerable reduction in capacity.

1. British Standards Institution, The Structural Use of Concrete, CP 110: Part 1, BSI, London, 1972.

2. Concrete Society, The Design of post-tensioned concrete flat slabs in buildings, Tech Report, The Concrete Society, London, 1974.

3. ACI-ASCE Committee 423, "Tentative recommendations for prestressed concrete flat plates", Journal of the American Concrete Institute Proceedings, Vol 71, No 2, Feb 1974.

19 Experimental investigations on the problem of lateral buckling of reinforced concrete beams

H. TWELMEIER and D. BRANDMANN
Technical University of Braunschweig, FRG

SUMMARY

Lateral buckling of reinforced concrete beams is related first of all to the decrease of bending and torsional stiffness in the beam due to cracking. Small-scale model tests show the influence of percentage of tensile reinforcement and of small transverse bending moments and torsional moments acting together with the main bending moment. The development of a suitable test set up and measuring technic is described. First pilot tests on lateral buckling confirm the results of stiffness tests.

1. INTRODUCTION

Due to the slenderness of wide-span prefabricated beams of reinforced concrete with I- or T-cross-section, the problem of lateral buckling did arise in concrete constructions too. Contrary to steel structures a mathematical solution of this problem is very difficult for beams of reinforced concrete, because the bending and torsional stiffness depend on the amount of loading and propagation of cracking and are variable along the longitudinal axis of the beam. Therefore in the last years several approaches have been made for approximate solutions. In most cases in the wellknown formulas for critical loads or moments given by the theory of elastic stability (see e.g. /1/) approximate values for reduced stiffnesses are inserted (e.g. /2, 3/). This doesn't correspond to real conditions very well, because due to unavoidable imperfections the problem of lateral buckling should be treated by taking into account the deformations of the beam ("2nd order stress theory"). However, in this case also safe assumptions are necessary for the stiffnesses involved. But the bending stiffnesses can be calculated rather exactly only when torsion is absent and the effective torsional stiffness is not known sufficiently at all, especially in the case of predominant bending here in discussion. So, due to the different simplifications made, the approximate solutions lead to different results, which could not be checked until yet because of the lack of suitable experimental test results.

For this reason an experimental program has been started at the Institute for Statics, Technical University of Braunschweig, which covers two main subjects:

1. An experimental investigation of the development of the bending stiffnesses (EI_y) and (EI_z) and the torsional stiffness (GI_T) in dependence on the bending moment M_y up to failure load. For these tests beam segments with rectangular and I-cross-section are used. Main parameter is the percentage of reinforcement. However, starting with pure bending, the effect of small transverse bending moments M_z and torsional moments M_T, which are acting at the same time and shall simulate the influence of goemetrical imperfections of real structures, is studied too. For notations see Fig. 3.

2. Several tests on small-scale reinforced concrete beams in order to study the mode of failure and the progress of lateral buckling. For comparison the values of significant parameters should be the same as in the stiffness tests.

2. TEST SPECIMENS

Due to the large dimensions of real beams which would tend to lateral buckling small-scale model tests are necessary. The dimensions given in Fig. 1 for beam segments with I-cross-section used in the stiffness tests are scaled down about 1 : 4 from common profiles used in prefabrication. Some additional tests have been carried out with reduced width of top and bottom chord and with beam segments of rectangular cross section b/d = 5/15 cm and 4/24 cm respectively.

Figure 1   Dimensions of profiled beam segments

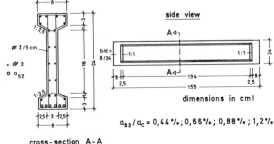

cross-section A-A

Corresponding to the model scale the concrete aggregate has been scaled down in the same ratio up to a maximum grain size of 4 mm following a grading curve in DIN 1045. Characteristic properties of the model concrete (mean values for prismatic specimens 4 x 4 x 16 cm) are a compression strength of 39 N/mm^2 and a direct tensile strength of 2,5 N/mm^2. From the observed cracking loads in the beam tests a tensile strength in bending of 4...5 N/mm^2 was calculated.

For the reinforcement multi-rib cold worked bars are used. For the main tensile reinforcement industrially fabricated bars of 4 and 5 mm dia. were available. For the other reinforcement including stirrups (see Fig. 1) adequate multi-rib bars were manufactured by rolling large deformations into the surface of plain round bars of 2 mm dia. using a special "profiling machine". All reinforcing bars have been brought to the quality of RK 420/500 by a controlled heat treatment up to 550° C.

3. TEST SET UP

3.1 General remarks

The test set up should be used for the stiffness tests and buckling tests in the same way as far as possible. This concerns first of all the conditions of support and loading. Additionally

the measurement of the stiffnesses had to be taken into account.
In the stiffness tests a constant moment M_y was applied to the
beam segments. So the bending stiffness (EI_y) could be derived
directly from the curvature $\varkappa_y$ occurring in loading. In order
to measure the transverse bending stiffness (EI_z) and the torsional stiffness (GI_T) present at the same time it seemed
favourable to apply small constant moments M_z and M_T ($= M_x$) at
each load level of main bending and to compute these stiffnesses by the help of the basic formulas of the theory of elasticity from the accompanying curvatures $\varkappa_z$ and lateral rotations φ_x. However, because these additional moments must not
increase cracking, their amount has to be kept very small but
simultaneously constant in order to avoid incorrect measurements. So the bearing construction and loading devices should
produce no friction or unplanned restraint and all loads
applied should not be influenced by the deformations of the
test specimens, neither in amount nor in direction. This is
valid also for those stiffness tests, in which the effect of
geometrical imperfections is simulated by small moments M_z and
M_T acting at the same time (see chapter 5) and for the tests on
lateral buckling. The schematic test set up is shown in Fig.2.

Figure 2   Schematic test set up

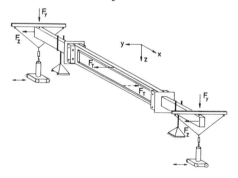

3.2 Conditions of support and bearing construction

The ideal support conditions representing a classical "fork-support" are given in Fig. 3. Big effort has been made in order
to fulfill these conditions. The solution gained step by step

was a rather complicated combination of roller and ball bearings (see Fig. 4). For this reason the bearing construction

Figure 3   Ideal support conditions

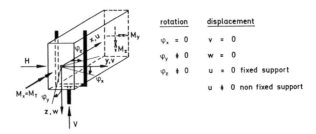

Figure 4   Bearing construction (schematic)

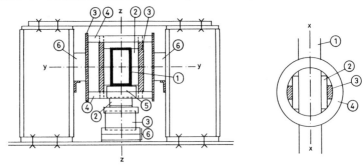

① steel cantilever
② axial grooved ball bearing
③ steel plate
④ grooved ball bearing
⑤ anti-friction bearing
⑥ roller bearing

was kept together during the stiffness tests. Each new test specimen was built in by the help of two steel cantilevers, which were part of the bearing construction (see Fig. 2).

3.3 Loading devices

The main bending load was applied by servo-controlled hydraulic jacks. In order to maintain the load direction vertical (see 3.1) the jacks are moved by the help of a motor-driven screwed spindle, when a displacement transducer indicated horizontal displacements of the point of application of load (see Fig. 5).

Figure 5 Loading device for main bending

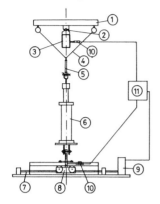

① loading beam
② ball and socket arrangement
③ test specimen
④ steel wire rope
⑤ load cell
⑥ hydraulic jack
⑦ screwed spindle
⑧ anchorage of the hydraulic jack (horizontal movable)
⑨ motor
⑩ displacement transducer
⑪ automatic control system

The amount of motion is controlled by a second displacement transducer positioned to the jack and an automatic control-system. The small additional moments M_z and M_T are applied by gravity loads F_z and F_T (see Fig.2), which are turned around in the horizontal direction by cables and guide pulleys.

4. MEASURING INSTRUMENTATION

The measuring program covered measuring of strains, displacements, curvatures, crack patterns and forces. For all deformations measurements displacement transducers are used. In order to measure strains they are inserted by the help of springs between anchore blocks cemented to the concrete surface. For the measurement of vertical and horizontal displacements twisting of the beam had to be taken into account. Therefore steel cables were connected to fixed measuring points at one end and - after leading them around guide pulleys - to fixed displacement transducers on the other end. So the accuracy of displacement measurement is not affected by the deformations of the beam: With the known length of the cables the real horizontal and vertical displacements and the rotation of the cross section can be calculated from the measured values. Due to the extensive measuring instrumentation shown in Fig. 6 with the possibility of calculating curvatures in different ways a lot of redundant measured values was available.This led to rather accurate mean values.

Figure 6   Measuring instrumentation for the stiffness tests

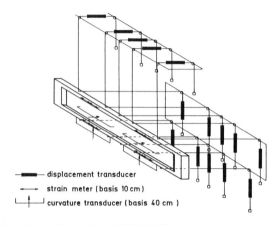

— displacement transducer
→ strain meter (basis 10 cm)
⌐┼┘ curvature transducer (basis 40 cm)

5. TEST PROGRAM AND TEST PROCEDURE

In order to take into account geometrical imperfections of real prefabricated beams (initial deformation v(x) of the compression chord and/or an initial inclination φ_x of the whole beam after erection) their effect was simulated in the stiffness tests by additional small moments M_z and M_T acting at the same time. The ratios of moments M_z/M_y and M_T/M_y are held constant when the main bending moment M_y was increased by the deformation controlled loading procedure described in chapter 3.3. At several levels of loading the measured values of all transducers were recorded by a computer controlled data logging system. All tests were finished when the ultimate moment M_{yu} was reached by yielding of the main tensile reinforcement a_{s2}.

6. FIRST TEST RESULTS

6.1 Stiffness tests

Some characteristic test results from the stiffness tests are given in Fig. 7. When torsion is absent, the curves for the bending stiffnesses consist of two nearly horizontal sections describing the phases of uncracked stage I and fully developed crack-pattern respectively and the decrease between. The length of the initial horizontal sections depends on the magnitude of

Figure 7 Load-dependent development of stiffnesses

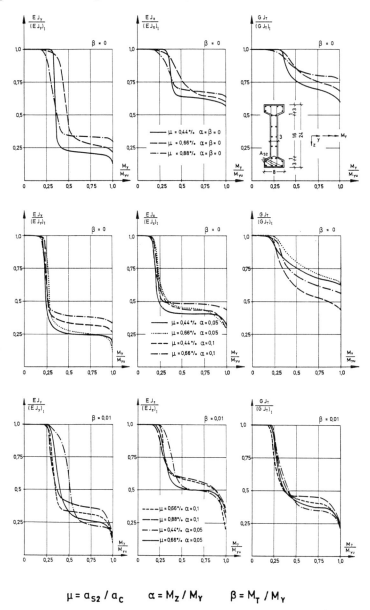

$$\mu = a_{S2} / a_c \qquad \alpha = M_Z / M_Y \qquad \beta = M_T / M_Y$$

tensile strength and the slope of the decrease due to cracking shows the effect of tension stiffening. The results obtained could be confirmed rather good by computed values. In the curves for the torsional stiffness the slope of decrease due to cracking is more inclinated. This shows the effects of aggregate interlock and dowel action of the reinforcement in the cracked region, which also result in considerable larger final values of (GI_T) than those calculated from approximate foumulas published. Additional torsion will lead to a more similar form of all curves. In all cases the stiffness $(EI)_z$ und (GI_T), which are most important for the occurance of lateral buckling, are affected by the additional moments M_z and M_T significantly.

6.2 Pilot tests on lateral buckling

The test beams with rectangular cross section (one of them with a pre-deformed compression zone) were 1 : 4 scale models of examples, which Röder /4/ had treated mathematically by the

Figure 8 Moment-displacement diagrams for lateral buckling

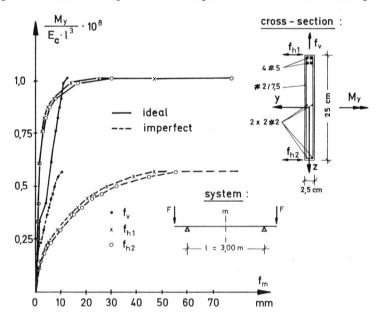

help of nonlinear theory of elasticity. The results given in
Fig. 8 show the typical failure mode - bending accompanied by
twisting - and the effect of geometrical imperfections. In this
case the initial deformation caused a rather large bending in
the transverse direction, so that the tensile reinforcement
started to yield earlier than in the "perfect" beam. In both
cases the experimental failure load was larger than the predicted one. This will be explained first of all by the fact,
that the torsional stiffness has been assumed too small in the
calculations.

7. CONCLUSIONS

In order to get safe results multiple checks have been performed. The behaviour of all components of the reinforced model
concrete used had been tested before in simple beam tests. The
test set up and the measuring technic have been proved for both
the stiffness tests and the buckling tests by the help of idealelastic beams out of aluminum and comparing calculations. After
finishing the stiffness tests, the next task will be to develop
simple models and formulas which approximate the measured results rather sufficient.

The investigations are sponsored by the Deutsche Forschungsgemeinschaft (DFG).

REFERENCES

/1/ TIMOSHENKO, St and GERE, J, Theory of Elastic Stability, McGraw-Hill Book Company, New York, (1961).

/2/ RAFLA, K, 'Vereinfachter Kippsicherheitsnachweis profilierter Stahlbetonbinder', Die Bautechnik, Vol. 50, No.5, (May 1973).

/3/ JELTSCH, W, Ein einfaches Näherungsverfahren zum Nachweis der Kippsicherheit von Stahl-, Stahlbeton- und Spannbetonträgern, Dissertation, TH Graz, (1970).

/4/ RÖDER, F K, Berechnung von Stahlbeton- und Spannbetonträgern nach Theorie II. Ordnung, Dissertation D 17, TH Darmstadt, (1982).

DISCUSSION

Dr F K Garas   Taylor Woodrow Construction Ltd   UK

Do you intend to introduce the effects of aggregate interlock and dowel action into the design formulas?

Author's Reply

In order to give a substantiated answer we must discuss the results a little bit more in detail. In addition to Fig 7 in the paper Fig 9 is given below where those values of the stiffnesses are compared, when the tension reinforcement starts to yield. The change in the bending stiffnesses is comparatively small and depends first of all on the position of the zero line for stresses, which rotates due to the action of M_z. Besides that the effective tensile strength, which is not equal for all beams, has some influence. So the effects of aggregate interlock and dowel action of the reinforcement are only important for the torsional stiffness, where they can alter the slope in the moment-stiffness-diagrams (Fig 7) as well as the final values (Fig 9). In order to decide if we can consider these effects in safe design formulas we must check by which parameters they will be influenced.

The effect of aggregate interlock depends on the maximum size of aggregate, on the crack width and on the acting combination of loading. In Fig 9 we see that transverse bending (M_z) reduces the values of (GJ_m) significantly because apart from a certain modification of the shape of the compression zone – the crack widths increase in a part of the beam. An even larger reduction results from the action of additional torsion.

Further on repeated loading and unloading in real structures due to the changes in live load will enlarge the crack widths and consequently reduce the effect of aggregate interlock.

Finally the reduced scale of the model beams must be considered. The maximum size of aggregate ($\emptyset$ 4 mm) has been scaled down and seems to be not too large in comparison with real beams. However, crack widths do not scale linearly in general and that portion of crack height with small width favourable for interlock is probably larger than in prototypes. So the effect of aggregate interlock deduced from small scale models may be overestimated.

Summing up the effect of aggregate interlock seems to be an incalculable quantity, which can help us to explain some results but should not be used in design formulas. At best this effect will give an additional margin of safe

It is another fact with the effect of dowel action. This effect depends on the crack width too and besides that on the precondition that the reinforcement does not yield. Because the main bending reinforcement, which will produce the effect of dowel action first of all, is situated in that region of the beam, where the crack widths are largest, it seems to be permissible to transfer model results on real structures. Further on lateral buckling will only influence the safety of the beam if it can occur before the ultimate design load for bending is reached. That means, that the reinforcement does not yield, so we will take into account the effect of dowel action in our design formulas. Some remaining uncertainties may be compensated taking a safety factor for lateral buckling which is somewhat increased compared to bending. The dotted line in Fig 9 shows that the increase of the torsional stiffness due to dowel action will be at least 50%. (The corresponding curves for α and/or $\beta \neq 0$ lie below).

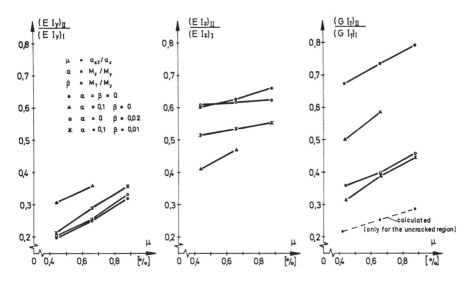

Fig.9: Stiffnesses at final crack-patterns for $M_y \rightarrow M_{yu}$

(Longitudinal reinforcement starts to yield)

20 Experimental behavior of thick pile caps

R. JIMENEZ-PEREZ
University of Puerto Rico, Mayaguez, Puerto Rico
G. SABNIS
Howard University, Washington, DC, USA
A. B. GOGATE
Ohio State University, Columbus, USA

SUMMARY

An experimental investigation into the behavior of thick pile caps is presented herein to determine the effect of the concrete compressive strength and shear span ratios to determine the ultimate vertical load sustained by the pile cap. The tests were conducted on model specimens with scale ratios of 5:1 and were compared with the results of larger scale models reported in the literature.

The experimental results obtained indicated that the ultimate load capacity is a strong function of the concrete tensile strength, increasing proportionally with the concrete strength. The amount of flexural reinforcement did not increase the ultimate load sustained by the pile cap as long as the reinforcement ratio exceeds the minimum amount prescribed for temperature considerations. A nondimensional analysis of the data showed that an analytical relationship could be established to predict the ultimate load capacity of thick pile caps once the concrete compressive strength, the number of piles and the pile diameter are known. The derived relationship is adequate to predict the ultimate load capacities of pile caps with symmetrical pile arrangements and with shear span ratios less than one.

INTRODUCTION:

The design of thick pile caps for shear presents a serious problem as the design provisions available were based on the behavior of structural elements that did not meet the geometric relations of pile caps as actually constructed. Thick pile caps may be defined as structural elements whose thickness is equal to or greater than the distance from the pile center line to the face of the supported column as shown in Fig. 1. For this elements, the shear span, defined by the ratio of the distance from the supported column to the pile center line, a, to the depth to the flexural reinforcement, d, is usually unity or smaller. As a matter of reference, the ACI Building Code (ACI-318-83) (1) does not provide any specific recommendations for the shear design of thick pile caps and designers are forced to assess the shear strength of this structural elements from the provision for the shear strength of deep beams. In this respect also, the designer is not illustrated as to where the critical sections for shear are located. Actually however, the behavior of thick pile caps is similar to that of a thick slab and its shear strength can not be established unless the tridimensional response of the element is adequately modeled.

The existing literature provides various analytical methods by which the load capacity of the pile cap can be determined if the loads are assumed to be distributed to the piles by a tridimensional truss(2, 3, 4, 5). The Concrete Reinforcing Steel Institute (6) has developed a methodology for the shear design of thick pile caps that assumes that the pile cap behaves as a two-way deep catilever slab with theoretical shear strengths of $0.5f_c'$ for pure shear conditions and $4\sqrt{f}$ for locations at d/2 around the column perimeter. These theoretical investigations have not been substantiated as the amount of reliable experimental data is limited to that provided by Clarke (7). Thus there is a strong need to conduct an experimental investigation into the behavior of thick pile caps in order to corroborate the analytical models and to determine the response of these elements.

An experimental investigation has been initiated, using small scale models, to determine the behavior of thick pile caps. The test program designed uses small scale models with scale factors of 1:5 approximately. The various parameters considered in this program are the concrete compressive strength, the shear span ratio and the amount of flexural steel provided. In this paper, the results obtained for the first 25 specimens with a shear span ratio ranging from 0.4 to 1.0 are reported.

EXPERIMENTAL DETAILS AND PROCEDURES:

The geometry of the pile cap model specimen used is shown in Fig. 2. The pile cap dimensions are 13 inches square with a total depth of 6 inches. The piles and columns were simulated by means of steel cylinders 3 inches in diameter. The concrete was provided by a ready-mixed concrete producer with a 3/8 in. maximum aggregate. The concrete compressive strengths for the specimens tested to date ranged from 2364 psi to 5409 psi.

The flexural steel was arranged in a grid pattern using the same reinforcement ratio in each direction. The reinforcement was fully anchored as shown in Fig. 2 to avoid any premature bond failure. The reinforcement ratios provided ranged from .0014 to .0101, while the nominal depths to the flexural reinforcement ranged from 4 to 4.75 inches.

The model specimens were tested in a universal testing machine with a 250,000 lbs. capacity once the steel cylinders used to model the column and piles were set with a plywood template. The compressive load was then applied to the specimens and readings where taken at various intervals until failure of the specimen occurred. Upon loading, the first cracks were observed at the bottom of the central portion of the vertical faces of the specimen. This cracks propagated in the vertical direction as the load was applied, and near failure the crack pattern shown schematically in Fig. 3 for specimen SS02 was typically observed. The crack pattern observed on the bottom of most of the pile caps evidenced a punching shear failure although various specimens failed by concrete splitting of the pile cap into two halves.

ANALYSIS OF TEST RESULTS:

The ultimate loads sustained by the model pile caps are summarized in Table 1 together with the shear span ratio, the depth to the flexural reinforcement, pile diameter, and the concrete strength of the specimen. In this table, the experimental results reported by Clarke (7) for the specimens that had full anchorage of the flexural reinforcement are also included. These tests were performed on specimens that were either 36 or 30 inches square with a total depth of 18 inches. All the specimens considered had the flexural reinforcement placed in a grid pattern with a yield strength of approximately 60 ksi. The 7.8 in. in diameter piles were casted monolithicaly with the pile cap together with the 8 inches square column. It should be mentioned that the concrete compressive strengths reported in Table 1 have been modified to reflect the concrete strengths of the standard 6 by 12 inch cylinders by modifying the values of the 3 by 6 inch cylinder results used in the model tests and the 6 inch cube strength reported by Clarke (7).

The results obtained to date indicate that the ultimate load sustained by the pile cap is strongly dependant on its concrete compressive strength as a measure of the tensile capacity of the concrete. The flexural reinforcement ratio provided is not critical as long as the minimum amount required for temperature considerations is present. The analysis of the available experimental data shown in Table 1 can be simplified if nondimensional terms are used. For this case a nondimensional strength parameter, given by the following equation can be related to the tensile capacity of the concrete, expressed as a function of the concrete compressive strength in nondimensionless terms also:

$$S = \frac{P_{exp}}{n d_p^2 f_c'} \qquad (1)$$

where: S = nondimensional strength parameter
P_{exp} = experimental load sustained by pile cap
n = number of piles in pile cap
f_c' = concrete compressive strength
d_p = pile diameter

In Fig. 4, the nondimensional strength parameter is plotted as a function of the square root of the concrete compressive strength for all the 32 specimens detailed in Table 1. It can be seen that the strength parameter decreases consistently with increasing values of the concrete strength. It is also evident that for any given concrete strength there exists a wide scatter of the data would be expected for any behavior that is strongly dependent on the tensile strength of the concrete. The spread of the experimental data at any given concrete strength is also influenced by the various shear span ratios used in the tests. However, at this time there is no adequate information to assess the effect of the shear span ratio on the shear strength of the pile cap.

The trend shown in Fig. 4 can be represented by the following linear relation:

$$S = -6.163E-3 \sqrt{f_c'} + 0.807 \quad (2)$$

The calculated failure loads determined from Eqn. 2 are tabulated also in Table 2 for all the tests considered together with the ratio of calculated to experimental failure loads. The linear regression to the experimental values is good as the failure load of approximately 84% of all the specimens shown is within 15% of the experimental load.

For design purposes however, it is convenient to use a lower bound to the experimental data shown in Fig. 4 to relate the nondimensional strength parameter to the tensile strength of the concrete, indicated in the graph by the solid line. The relation proposed as the lower bound is given by the following relations:

$$2000 < f_c' \leq 5000 \text{ psi} \quad S = \frac{(109.7 - \sqrt{F_c'})}{130} \quad (3a)$$

$$f_c' > 5000 \text{ psi} \quad S = 0.3 \quad (3b)$$

The calculated failure loads determined from Eqn. 3 are also compared in Table 2 to the experimental failure loads where it can be concluded that the lower bound formulation predicts loads that are lower than the experimental loads for approximately 84% of the specimens. The smallest ratio of calculated to experimental load is 0.72.

The ultimate vertical load sustained by thick pile caps with a symmetric arrangement of piles with respect to its geometric center, and with shear span ratios smaller than 1.0, can be determined from Eqn. 3 once the number of piles, the pile diameter and the concrete strength has been selected. The experimental data shown in Fig. 4 can also be interpreted as the unit shear stress sustained by the pile cap per pile or as the punching shear capacity of the thick slab. In this context and for concrete strengths from 3000 to 6000 psi, the punching shear strength ranges from 42% to 30% of the concrete compressive strength. In terms of the square root of the concrete compressive strength, the maximum punching shear stress ranges from 21.2 to 23.2 times the square root of the compressive strength. It can be concluded from this analysis that the current punching shear strength of 4 times the square root of the concrete compressive strength provided by present design regulations is extremely conservative for the design of thick pile caps.

CONCLUDING REMARKS:

The experimental investigation reported herein has shown the adequacy of small scale models to represent adequately the behavior of thick pile caps when compared to larger prototypes tested by other investigators. The techniques of model analysis have proved useful to determine the ultimate load capacity and the punching shear strength of thick pile caps and to determine a mathematical relation that can be used to predict confidently their load capacities. Additional model tests are underway to determine the effect of the shear span ratio, the arrangement of the flexural reinforcement, and the scale ratio of the specimens. With these additional tests a more comprehensive analytical relation to predict the ultimate loads can be developed.

REFERENCES:

(1) ACI Committee 318, 'Building Code Requirements for Reinforce Concrete (318-83)', American Concrete Institute, Detroit, Michigan, (1983)

(2) BLEVOT, J and FREMY, R, sur Pieux', Annales, Institute Technique du Batiment et des Travaux Publics(Paris), V.20, (Feb. 1967).

(3) WHITTLE, R T, BEATTLE, D, 'Standard Pile Caps', Concrete, (January 1972).

(4) YAN, H T, 'The Design of Pile Caps', Civil Engineering and Public Works Review, Vol. 49, (May and June 1954).

(5) HOBBS, N B, STEIN, P, 'An Investigation into the Stress Distribution in Pile Caps with some Notes on Design', Proceedings of the Institution of Civil Engineers, Vol. 7, (June 1957).

(6) CRSI Handbook, 3rd edition, Concrete Reinforcing Steel Institute, Chicago, (1978).

(7) CLARKE, J L, 'Behavior and Design of Pile Caps with Four Piles', Technical Report No. 42.489, Cement and Concrete Association, Wexhan Springs, (Nov. 1973).

(8) SABNIS, G, and GOGATE, A B, 'Investigation of Thick Slab (Pile Cap) Behavior', American Concrete Institute Journal, Title 81-5, (Jan-Feb 1984).

SPEC NO	NOMINAL DEPTH (d)	SHEAR SPAN RATIO	PILE DIA. (inches)	CONC STRENGTH (psi)	PEXP (Kips)
SS01	4.3875	0.95	3.0000	4124.00	56.30
SS02	4.3945	0.95	3.0000	4124.00	55.00
SS03	4.3650	0.95	3.0000	4124.00	55.75
SS04	4.3945	0.94	3.0000	4124.00	50.75
SS05	4.2750	0.97	3.0000	5409.00	59.25
SS06	4.2750	0.97	3.0000	5409.00	63.00
SG02	4.625	0.90	3.0000	2364.00	39.00
SG03	4.625	0.90	3.0000	2364.00	39.75
MS01	4.5	0.92	3.0000	4160.00	62.00
MS02	4.5	0.92	3.0000	4160.00	62.00
MS03	4.5	0.92	3.0000	4160.00	69.00
MS04	4.75	0.88	3.0000	4160.00	65.50
MS05	4.75	0.88	3.0000	4575.00	52.00
MS06	4.25	0.98	3.0000	4160.00	58.75
MS07	4.25	0.98	3.0000	4160.00	64.75
MS15	4.625	0.90	3.0000	4575.00	67.50
MS16	4.626	0.90	3.0000	4575.00	65.00
MS17	4.50	0.92	3.0000	4575.00	69.75
MS19	4.5	0.92	3.0000	4575.00	72.00
MS20	4.25	0.98	3.0000	4575.00	69.75
MS23	4.25	0.98	3.0000	4575.00	70.50
MS24	4.25	0.98	3.0000	4575.00	74.50
MS28	4.00	1.04	3.0000	4160.00	71.75
MS29	4.00	1.04	3.0000	4160.00	66.00
MS30	4.00	1.04	3.0000	4575.00	70.50
A007	16.14	0.76	7.8740	3346.00	247.19
A010	16.14	0.76	7.8740	2896.00	342.00
A011	16.14	0.76	7.8740	2750.00	369.00
A012	16.14	0.76	7.8740	4074.00	369.00
B001	16.14	0.41	7.8740	4336.00	468.00
B002	16.14	0.41	7.8740	3958.00	421.00
B003	16.14	0.41	7.8740	5834.00	398.25

TABLE 1: Summary of Test Results

		EQUATION 2		EQUATION 3	
SPEC NO	PEXP	PCAL (kips)	RATIO Pcal/Pexp	PCAL (kips)	RATIO Pcal/Pexp
SS01	56.30	61.26	1.0881	51.94	0.9226
SS02	55.00	61.26	1.1138	51.94	0.9444
SS03	55.75	61.26	1.0989	51.94	0.9317
SS04	50.75	61.26	1.2071	51.94	1.0235
SS05	59.25	69.19	1.1678	58.42	0.9859
SS06	63.00	69.19	1.0983	58.42	0.9273
SG02	39.00	43.27	1.1095	39.99	1.0253
SG03	39.75	43.27	1.0886	39.99	1.0059
MS01	62.00	61.54	0.9926	52.07	0.8399
MS02	62.00	61.54	0.9926	52.07	0.8399
MS03	69.00	61.54	0.8919	52.07	0.7547
MS04	65.50	61.54	0.9395	52.07	0.7950
MS05	52.00	64.50	1.2404	53.29	1.0248
MS06	58.75	61.54	1.0475	52.07	0.8863
MS07	64.75	61.54	0.9504	52.07	0.8042
MS15	67.50	64.50	0.9556	53.29	0.7895
MS16	65.00	64.50	0.9923	53.29	0.8198
MS17	69.75	64.50	0.9247	53.29	0.7640
MS19	72.00	64.50	0.8958	53.29	0.7401
MS20	69.75	64.50	0.9247	53.29	0.7640
MS23	70.50	64.50	0.9149	53.29	0.7559
MS24	74.50	64.50	0.8658	53.29	0.7153
MS28	71.75	61.54	0.8577	52.07	0.7258
MS29	66.00	61.54	0.9324	52.07	0.7890
MS30	70.50	64.50	0.9149	53.20	0.7559
A007	247.19	374.90	1.5166	331.00	1.3390
A010	342.00	342.26	1.0008	308.75	0.9028
A011	369.00	330.76	0.8964	300.39	0.8141
A012	369.00	419.33	1.1364	356.51	0.9662
B001	468.00	432.95	0.9251	362.73	0.7751
B002	421.00	412.91	0.9808	353.27	0.8391
B003	398.25	488.92	1.2277	434.05	1.0899

TABLE 2: Summary of Experimental and Predicted Loads

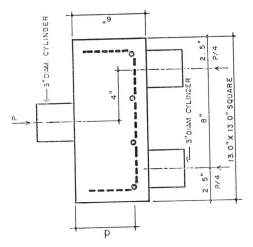

Figure 2 Geometry of pile cap test specimen

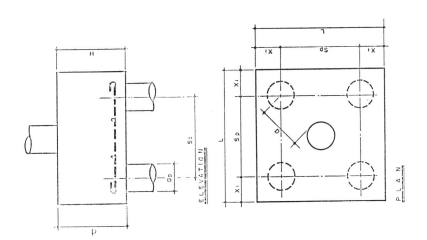

Figure 1 Geometry of thick pile caps

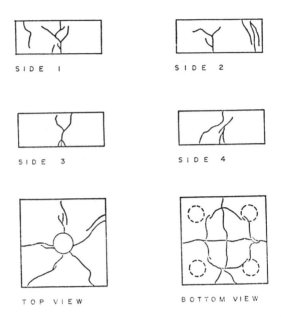

Figure 3 Crack pattern at failure for typical specimen

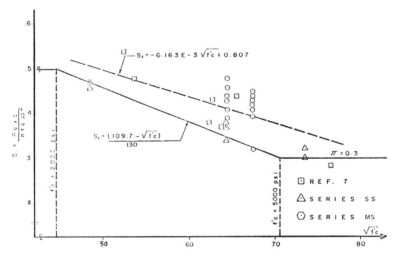

Figure 4 Nondimensional strength parameter vs. concrete tensile strength

DISCUSSION

C D Goode   University of Manchester   UK

The authors mention that the a/d ratio is a significant parameter yet Equation 1 and Figure 4 do not include it. I consider that some factor, such as the 2d/a included in CP 110, should be included when shear spans are short to give the equation more general applicability.

21 Model analysis in impact research

J. EIBL and M. FEYERABEND
University of Karlsruhe, FRG

1 Problem

In case of structural problems which are to be studied by experiments, rather small models are often used to reduce the costs of test specimens and test equipment. This type of research rises the question of similarity and apropriate transfer laws.

A slightly different situation is given when the size of the structure to be tested is so big - e.g. in case of nuclear power plants - that even rather big test specimens still have to be regarded as models. This type of model testing promotes experiments in combination with modern numerical analysis. Here the model is used to justify e.g. a rather sophisticated theoretical Finite Element approach which otherwise could not be trusted. Having reached a state of knowledge where the numerical simulation can describe models of different size correctly one has a universal tool to predict the behaviour of the real structure. Thus the problem of similarity is circumvented.

During the last ten years the author gathered some experience with this latter type of model analysis studying impact problems, on which a short report will be given.

2 Experiments

During the last seven years impact tests on reinforced concrete models have been carried out including

- beams,
- plates, and
- columns.

In all cases the test specimens have been loaded by hard impact, which to the author's own definition is characterized by the limiting case of rigid mass striking the structure to be studied. Neglecting secondary losses the full kinetic energy is transfered to deformation energy of the struck body.
These conditions were approximately reached in experiments described as follows (fig 1):

In a vertically orientated steel tube a cylindrical body is accelerated by gravity and additional air pressure applied at the top of the tube. At the bottom of the dropping weight strikers of different geometry - cone, hollow cylinder etc. - may be attached. The latter impinging on the concrete structure at the lower end of the tube may be assumed to be rigid in comparison with the deformation of the struck structure. Contact forces are simultaneously measured by a load cell between dropping weight and striking head, an accelerometer at the falling mass, and a special electronical device controlling the speed of the falling mass.

2.1 Beam Tests

Beam tests were carried out on altogether 30 differently reinforced concrete beams mainly 0.30 x 0.40 x 4.00 [m] in dimension (fig 2). While details are given in [1], [2], [3], the main experimental findings may be sketched as follows.
At the investigated speed of 1 - 20 [m/s] shear failures occured, the first through-going inclined cracks appearing a few milliseconds after contact of mass and structure. Much later secondary bending demages occured in the remaining parts of the structure if the beam was not completely destroyed in the first phase.
The shear cracks are the steeper the higher the applied speed was.
A principle explanation resulting from experiments - for detailed numerical studies see chapter 3 - may be given as follows:

If according to fig. 3 in a first approximation the inertial force p is linearly distributed according to the expected deflection, its intensity may be written as

$$R = \int_0^L p\,dx = \alpha \cdot F ,$$

α giving the percentage of the applied load.

Then for shear force and bending moment may be easily derived:

(2.1) $\quad M_{max}\big|_{x=\frac{L}{2}} = \frac{F \cdot L}{4}[1 - \frac{2}{3}\alpha]$

(2.2) $\quad Q_{(x)} = \frac{F}{2}[1 - \alpha + 4\alpha\frac{x^2}{L^2}]$

(2.3) $\quad Q\big|_{x=\frac{L}{2}} = \frac{F}{2}$.

For the limiting cases $\alpha = 0$ - static case - and a fictious case $\alpha = 1$ - all loads F carried by inertial forces, no reaction at the supports - the shear force remains

$$Q\big|_{x=\frac{L}{2}} = \frac{F}{2} ,$$

while the bending moment is reduced to one third of the static case. More detailed investigations support this rather simple explanation.
Thus with growing inertial forces in a simple supported beam the shear force increases compared to the midspan moment. Shear failures at midspan predominate bending failures, leading to inclined cracks.

2.2 Plates

Altogether 20 plates mainly with the dimension 4.00 x 4.00 [m], and a thickness of 0.18 - 0.40 [m] have been tested. The intention was to study the behaviour of different types of steel in plates subjected to bending and especially the effectiveness of shear reinforcement in form of stirrups (fig. 4).
The findings may be sketched as follows:

Under the applied speed of 3 - 25 [m/s] shear or bending failure occured depending on the amount of shear reinforcement. Thus it was shown that appropriate shear reinforcement may avoid a primary shear failure. Local strain concentration and the maximum endurable strain controlls failure. High local strain concentration occurs with high bond quality of the steel.

Having the same bearing capacity in tension lower grade steel with poor
bond behaves better then a high steel quality with good bond.
The reasons for unexpected shear failures were the same as in the case of
beams as was further verified by the methods discussed in chapter 3.

2.3 Columns

In a first phase vertically orientated columns with a heavy mass on top
of the column were struck by a horizontally accelerated mass (fig. 5).
The aim was to study the interaction between longitudinal force, bending
and shear forces in bridge piers or buildings under vehicle impact.
In a second phase - now under progress -, because of greater simplicity
in testing the column was horizontally orientated and vertically struck.
In this case the mass, on top of the real column under consideration, was
set on movable bearings at one end of the lying test specimen and
prestressed against the other end by a force

$$F = m \cdot g$$

(fig 6, 7).
The principle result may be explained as follows (fig 8):
During a first horizontal deflection the end of the column moves
downwards with high speed, thus reducing the contact force between column
and mass on top. Having reached the maximum deflection the horizontal
deformation reverses as well as the longitudinal. Now the end of the
column moves upwards against the meanwhile accelerated mass, thus
generally resulting in a contact force which is greater than the original
static force due to gravity.

3 Computations

According to the aims formulated in chapter 1 in all cases at least an
approximated quantitative description of the behaviour watched in the
experiments was tried.

3.1 Beams

The beam tests were modelled as shown in fig. 9, so that they could be easily treated by Finite-Difference-Methods (F.D.) as an initial value problem. This can be done even by normal electronic desk calculators. For a good agreement with the carried out tests the deformation characteristics of the spring k

$$k = f(w_0 - w_j)$$

simulating the contact force during the process of impinging is important. Hints how to treat this problem are given in [1] and [4], further studies being under progress at the author's institute and by some other researchers.

For engineering design purposes also the two mass model of fig. 10 proved to be very useful because of the simplicity.

3.2 Plates

In the meantime a computer program has been developed, by which the governing initial value problem at a rotational thick plate is solved. Time is discretized by Finite-Difference, space by Finite-Elements. The program includes a triaxial constitutive law for concrete, simplified strainrate effects, crack opening and of course bending and shear reinforcement (fig. 11). As this program was especially developed for the purpose of investigating airplane-crashes on structures it starts with a given force F(t) resulting from soft impact, where the force is practically not influenced by the deformation of the concrete structure. All kinetic energy is transfered to deformation energy of the striking plane structure.

For design purposes also in this case a two mass model proved to be satisfactory having the advantage of a simple numerical algorithm as in principle discussed in [5].

The resistance of the plate $F_1 = f(w_1)$ observes thorough attention (fig. 12).

Phase 1 in fig. 12 gives the straining of the uncracked concrete, phase 2
the following crackformation and straining of the vertical stirrups and
phase 3 finally the activation of the longitudinal bending reinforcement.
Herewith it may be explained

- that the given maximum force F_m sometimes causes only small
  damage $(w < w_1^{①})$

- that in other cases with scarcely different F_m due to the scattering
  tensile strength the same F_m causes great damage with $w_1^{①} < w_1 < w_1^{②}$ or
  even complete failure if there is not sufficient steel elongation
  available

- that some authors incorrectly conclude from their test results with
  "underreinforced" structures, that reinforcement is of no influence on
  penetration e.g.

Nevertheless realistic results are only found if also the bending
stiffness of the remaining plate around the shearcone is considered in
the model by

$$F_2 = f(w_2).$$

3.3 Columns

The tested column structures as described in chapter 2.3 have been
modelled as beams similar to fig. 9 just including the longitudinal
action as described in [1].
Because of the nonlinear deformation characteristics in the contact zone
of the reinforced concrete beam resp. steel beam the differential
operators had to be substituted by an appropriate difference scheme and
numerically to be solved.

4 Conclusions

The test of reinforced concrete models which are still large in spite of a scale factor applied and a description of the watched behaviour indepent of the tested dimensions may help to overcome the problem of similarity. Such a procedure may be classified as a special typ of model analysis.

Examples for this method are given treating structures under impact loads such as

- beams,
- plates, and
- columns with longitudinal and transversial loads interacting.

The results also give some general insight in structures under impact loading.

[1] Block, K.   Der harte Querstoß - Impact - auf Balken aus Stahl, Holz und Stahlbeton
Dissertation, Dortmund 1983

[2] Eibl, J.
Block, K.
Stoßbeanspruchung von Stützen durch Fahrzeuge und andere Stoßlasten
Universität Dortmund, Abteilung Bauwesen, 1980
Abschlußbericht B II 5 - 81 07 05-213/1
Bundeminister für Raumordnung, Bauwesen und Städtebau

[3] Eibl, J.
Block, K.
Zur Beanspruchung von Balken und Stützen bei hartem Stoß (Impact)
Bauingenieur 56 (1981), S. 368-377

[4] Nilson, L.   Finite Element Analysis of Impact on Concrete Structures. F.E. in Nonlinear-Mech. Vol. 2, Norwegian Institute of Technology, Trondheim, 1977

[5] Eibl, J.   Stahlbetonkonstruktionen unter Stoßbeanspruchungen, in: VDI Berichte 355, Der Sicherheitsnachweis für mechanische Strukturen unter transienter Belastung, Kolloquium Darmstadt, 1979

Fig. 1 Impact generating equipment

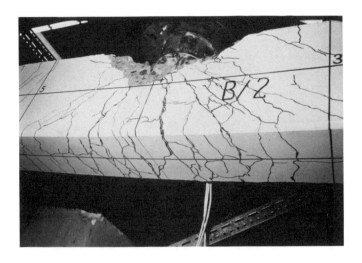

Fig. 2 Underside view on concrete beam after application of impact loading

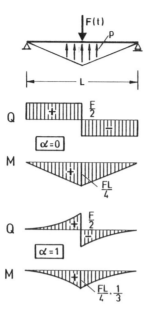

Fig. 3 Distribution of inertia forces analogside the beam; moments and shear forces for extreme values α

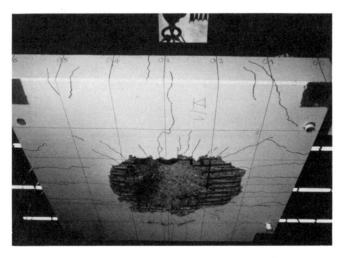

Fig. 4 Underside view on concrete plate after application of impact loading

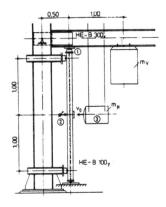

Fig. 5 Vertically loaded column under horizontal impact

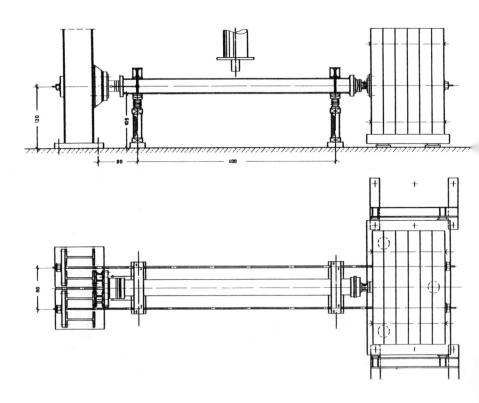

Fig. 6 Testing equipment for studying vertically loaded columns under horizontal impact

Fig. 7 "Vertically" loaded steel column before testing

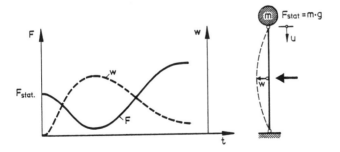

Fig. 8 Vertically loaded column under horizontal impact, principle result

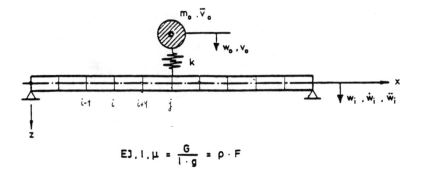

Fig. 9 Model for beam analysis

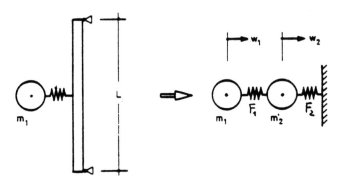

Fig. 10 Mechanical model for approximate analysis of an impact loaded beam

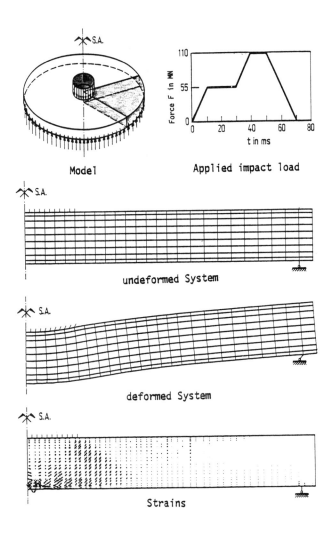

Fig. 11 Analysis of an impact loaded plate

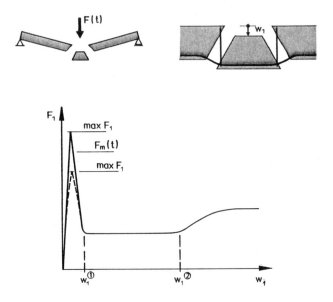

Fig. 12 Resistance function of an impact loaded plate with stirrups and longitudinal bending reinforcement

DISCUSSION

Dr Jorgen Nielsen   Technical University of Denmark

In formulating the problem it is stated in the paper that for very big structures it is a special advantage to use modern numerical analysis justified by tests to reach the state of knowledge, where the numerical simulation can describe structures of different sizes correctly. The authors claim that then a universal tool is thus obtained for predicting the behaviour of real structures, and the problem of similarity is circumvented.

In the following these fundamental considerations are discussed. One point is that the described method is a universal method with no special reference to very big structures. Another point is that if one succeeds in solving the problem as described, the problem of similarity is not circumvented but solved.

As a basis of the discussion an attempt is made to show that solving the problem of similarity and making a programme for correct numerical calculations are two sides of the same question. Figure 1 shows two ways of solving a given problem. As can be seen it is argued that the basic considerations are common. The only difference is that for the numerical analysis the fundamental equations have to be solved, which is much more complicated than forming a model law covering the same phenomena. Only phenomena considered are covered and if other phenomena play a role, errors will be seen in both cases.

The computer programme has also to be verified by tests, but these tests can be carried out more freely than tests according to a model law. The following remarks imply that the computer results are not empirically corrected by these tests. If development of the model law leads to contradictory requirements, scale errors can be foreseen but not determined, whereas they are included in the computer programme, which then predict the size effect. When verified it is normally much easier to get results from the computer programme than from the model law, because a new model has to be built.

The conclusion is, that it is a much bigger job to develop a computer programme than a model law, but a much better instrument if all phenomena in question are covered, ie the programme can be verified by tests.

In the cases where an advanced computer programme can be made it thus forms a tool to get more benefit from experimental results covering also what is usually called scale effects.

This is valid independent of the size of the structure, but it cannot justify an extrapolation as stated by the authors. The conference gave several examples of serious scale errors, which could be referred to the fact that new phenomena appeared in very small scales, and it is difficult to see why this could not be the case in very big scales. The problem of very big structures, which cannot be tested, is therefore not solved.

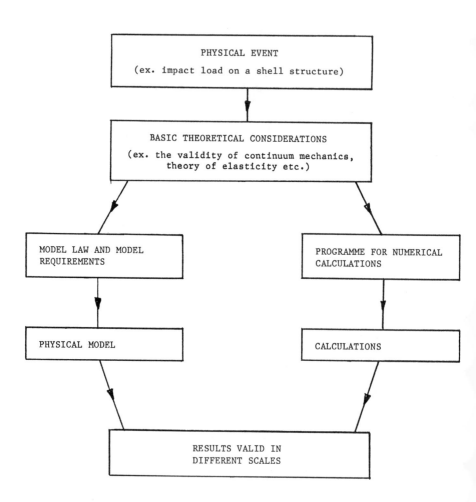

Figure 1. To a certain extent results from physical models and numerical calculations are based on the same considerations.

22 Modelling of reinforced concrete containment structures

PIOTR D. MONCARZ, JOHN D. OSTERAAS and
ANNE M. CURZON
Failure Analysis Associates, Palo Alto, California, USA

ABSTRACT

The use of experimental model tests to analyze the behavior of nuclear power plant containment structures is being explored. The primary purpose of such tests is to characterize containment structure integrity and rate of leakage resulting from over-pressurization representative of a severe nuclear accident. The nature of leakage and the complexity of the structure, built of reinforced concrete with an integral steel liner, put the problem beyond the limits of current analytical techniques. Properly designed scale model tests provide a feasible and reliable means of evaluating containment response in the post-elastic range. For the model test results to be acceptable, the model must be designed and constructed to ensure model-prototype similitude. Modelling concerns unique to the subject structure are presented along with suggestions for prototype-scale subassembly tests to aid in model development and response calibration. The selection of a model scale is also discussed in light of concerns such as economics, material availability, sophistication of labor, and acceptance by the engineering profession.

INTRODUCTION

One of the primary purposes of a secondary containment structure is to protect the environment from uncontrolled release of radioactive fluids during an accidental buildup of pressure and temperature inside the containment. This requires that the structural integrity of the containment is maintained, and that the rate of leakage through cracks in the structure is within acceptable levels even for the most severe accident conditions. In the United States and Canada, several extensive research programs have assessed the response of secondary containment structures in the post-elastic range up to failure. This paper results from an ongoing research program at Sandia National Laboratories in Albuquerque, New Mexico (1), which includes scale model and subassembly tests, as well as evaluation of existing analytical methods.

Despite advancements in analytical capabilities, post-elastic cracking and structural response of reinforced concrete structures cannot yet be reliably predicted. Model studies provide opportunities to investigate structural response and to accumulate experimental data for development of new analytical techniques and for reliability evaluation of existing techniques.

Techniques and parameters for strength modelling of reinforced concrete structures have been studied extensively and generally provide reliable results. Development of a model to accurately represent both strength and leakage response in the post-elastic range raises significant questions that are not readily addressed by available modelling experience. Similitude of cracking, welding scale effects, liner/concrete interaction, and local discontinuities all must be accounted for. This paper addresses those aspects of modelling unique to containment structures where integrity of strength and leak-tightness are studied.

PROTOTYPE DEFINITION

One objective of the containment integrity program is the study of post-elastic response to internal pressurization of a structure representative of the majority of reinforced concrete containment structures. Thus, the first step requires definition of such a generic structure. Based on a survey of existing and planned nuclear power plants, the prototype structure shown in Fig. 1 was developed. For economy the prototype included only those design and construction details which significantly affected the response parameters being investigated. The structure consists of a right cylindrical shell with a hemispherical dome, a flat base mat, and an internal steel plate liner. The steel liner is attached with headed shear

Figure 1. Generic containment structure geometry.

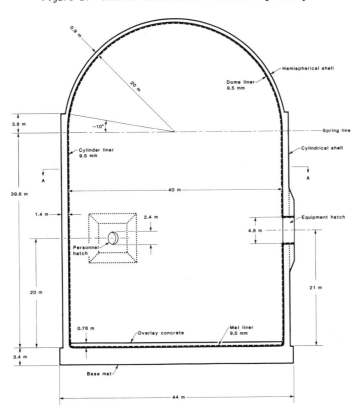

studs spaced 0.30 m (12 inches) on center. Reinforcement consists of orthogonal curtains of 57 mm (No. 18) bars on both the inner and outer faces of the concrete elements and diagonal seismic reinforcement at mid-thickness of the wall. Two major penetrations, the equipment hatch and the personnel hatch, and a cluster of major pipe penetrations are included in the model.

MODELLING CONSIDERATIONS

The prediction of prototype response from the results of scale model study requires that modelling laws be obeyed in the model design, construction, and testing processes. Besides the always-present test result variability, scale effects will influence the response and should be addressed.

Scale Selection

Selection of a model scale is subject to numerous considerations related to the test objectives, economical and physical test feasibility, result reliability, and acceptability of test results by the profession. Certain limitations narrow the range of feasible model scales available. Time constraints and physical limitations of the facilities provide the upper limit on the model scale. Construction techniques and availability of materials influence the lower limit of the scale range. A tradeoff occurs, however, when consideration is given to test acceptance by the engineering profession. Concern for model similitude prompts engineers to give greater credence to larger scale tests. Thus, in the final selection of a model scale, the relative values of economic savings and professional acceptance must be weighed.

The cost of a model is mainly a function of material cost, labor intensity, sophistication of labor force required, and technological requirements of the construction processes. Material cost and labor cost could each be represented as a step function with a major step between off-the-shelf material availability and conventional construction techniques for large models versus custom-made materials and advanced fabrication technologies for smaller models. In the containment model study, the main step occurs at about a scale of 1:9, as shown in Fig. 2. The change from commercial reinforcement to custom reinforcement presents the most significant step on this figure, as shown at the 1:9 scale value. For larger scales the material cost varies almost proportionally to the volume. The construction costs show a smaller scale-dependence. Small scale modelling techniques and custom-fabricated materials could be required for models smaller than 1:9. Such factors offset the economical benefits of the easier handling of smaller elements and reduction in material volume. Thus, within each scale range, the smallest feasible model provides the most economical solution.

Figure 2. Conceptual relationship between model scale and construction cost.

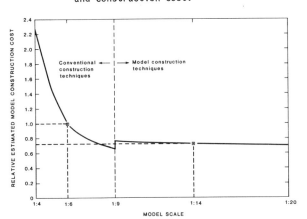

Similitude Requirements and Distortions

Based on similitude theory, a complete set of condition functions (scaling laws) can be determined to define model-prototype correspondence. Since not all similitude requirements can be identified and fulfilled, violations of some scaling laws have to be permissible. The researcher should ensure that, despite these simplifications, the model adequately represents the prototype.

Through-wall leakage requires both containment liner and concrete wall cracking. Liner strains and failure depend upon size and spacing of cracks in the concrete wall. Experience in the modelling of reinforced concrete structures shows that models tend to have fewer cracks than prototype structures. This implies wider cracks for properly scaled displacements. The data of five different researchers (2), all modelling the same prototype structure with only minor changes but at different scales, are presented in Fig. 3 to indicate the relationship between the nature of the crack spacing and scale. These data imply that the model can be expected to yield approximately one-half the number of cracks expected in the prototype structure under similar loading conditions. Based on these observations, it is necessary to experimentally establish correlation between leakage rate and crack sizes.

Figure 3. Number of cracks as a function of model scale.

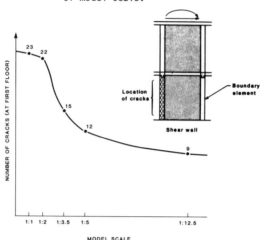

Similitude distortion in crack distribution introduces a major question related to the spacing of the liner anchors. Localized liner stretching results from liner spanning across concrete cracks. Therefore, with wider cracks, larger spacing between liner anchors should be considered. The appropriate spacing can be determined based on supporting test data. Thus, to minimize the similitude distortions in liner strains, the design of the model could require parametric study on structural subassemblies with different anchor spacing.

In reinforced concrete modelling studies, material simulation requires particular attention. Material properties representative of a prototype structure should be modelled. Thus, the strength parameters of the concrete should be based on aged prototype concrete (5 years), and the steel strength (reinforcement, liner) should correspond to the statistical distribution of the strength of prototype steel. It is reported that for lower strength concretes (3), the 25-year strength approached 240% of the 28-day strength. Tests of reinforcing steel showed a mean yield strength 18% higher than the specified value (4). In the modelling of massive structures, the prototype construction sequence should be reproduced.

TEST PLANNING

Containment integrity study requires an integrated experimental and analytical program such as shown schematically in Fig. 4. Certain critical elements such as liner welds, flaws in the liner, and joints and gaskets around penetrations can only be studied at the prototype scale utilizing subassembly tests. The distortions and scale effects introduced in the modelling can be determined through comparison of responses of prototype-size and model-size subassembies subjected to the same loading. This permits the development of a single data base, including both model- and prototype-scale test results. Such results may be used as a basis for the development of analytical prediction methods.

Figure 4. Interaction between model scale and prototype scale experimental studies.

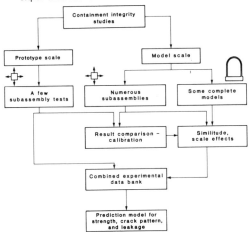

The nature of pressure tests (load control) does not allow for direct monitoring of crack development. Also, due to creep and relaxation effects, interruption of the test is undesirable. Thus, response parameters of interest must be monitored remotely.

SUBASSEMBLY TESTS

Calibrated crack leak tests provide a means of translating model-leakage data to prototype-leakage response including implicitly the scale effects on crack characteristics. Such tests also provide the data necessary for development of analytical techniques for leak rate prediction. Wall sections could be prepared and precracked and provided with a means of varying the crack width from test to test (5). Such testing may be most readily accomplished after the model testing is completed, with the model cracks being used as a guide to the development of the precracked test specimens.

Liner/concrete interaction for various anchor systems can be addressed through a series of subassembly tests. A test program currently under way to test prototype-scale wall panels (6) provides an opportunity for model-size specimen calibration. Liner response will vary with anchorage systems. Cracking response of the wall could also be affected by the anchorage system. Subassembly tests provide a feasible means to evaluate those effects.

Significant details omitted in the simplified prototype may also be studied through subassembly tests. Many leakage paths are possible through penetrations, the intricacy of which requires study on full-scale subassemblies. Leakage data from such tests may be combined with leakage data from the containment test to determine integrated leak rate under any loading conditions.

CONCLUSIONS

The definition of a structural prototype representative of an entire group of structures requires assembly and review of data on individual structures. This includes the geometry, reinforcement patterns, and material properties at the age of an "average" prototype.

The selection of a model scale is a complex process; considerations include material availability, construction techniques, physical parameters of testing facilities, reliability and believability of test results, and cost parameters. The cost change with model scale presents discontinuities at scales requiring change from off-the-shelf materials to custom-made materials, and from traditional to modelling construction techniques. The feasible scale range was found to be between 1:6 and 1:16; the upper bound (1:6) was the smallest model in which reinforcement commercially available in the United States could be used, and the lower bound (1:16) was the smallest model for which anchoring studs could be used feasibly. Financial and physical feasibility allowed the selection of the 1:6 model scale.

Scale effects and similitude distortions should be evaluated through subassembly support tests. Liner anchor spacing in the model should be selected based on comparison between crack parameters in the prototype-size and model-size subassembly specimens. The leakage rate through individual cracks can be studied on precracked subassembly specimens. A complete quality control and evaluation program is necessary to provide reliable information on the model material properties.

ACKNOWLEDGEMENTS

The research which led to many of the reported ideas is part of the project on Safety Margin of Containments carried out by Sandia National Laboratories under the sponsorship of the Nuclear Regulatory Commission (NRC). The authors are grateful to those institutions for the financial support that made this project possible.

REFERENCES

(1) MONCARZ, PIOTR D, OSTERAAS, JOHN D, CURZON, ANNE M, Experimental Modeling Techniques for Reinforced Concrete Containment Structures, Failure Analysis Associates Report No. FaAA 83-11-11 (March 1984).

(2) Proceedings of the Fourth Joint Technical Coordinating Committee, U.S.-Japan Cooperative Research Program Utilizing Large Scale Testing Facilities, Tsukuba, Japan (June 1983).

(3) WASHA, A W, and WENDT, K F, 'Fifty Year Properties of Concrete,' ACI Journal (January 1975).

(4) BUTLER, T A, and FUGELSO, L E, Response of the Zion and Indian Point Containment Buildings to Severe Accident Pressures, NUREG/CR-2569, LA-9301-MS, Los Alamos National Laboratory, Los Alamos, New Mexico (May 1982).

(5) RIZKALLA, S H, LAU, B L, and SIMMONDS, S H, 'Air Leakage Characteristics in Reinforced Concrete,' Journal of the Structural Division, ASCE, Vol. 110, No. 5 (May 1984).

(6) JULIEN, J T, WEINMANN, T L, and SCHULTZ, D M, Concrete Containment Structural Element Tests, Phase I, Final Report, Construction Technology Laboratories, Skokie, Illinois (March 1984).

Author's Closure

The reliability of the analytical method is a major question when studying a nuclear containment structure to attempt establishing its safety margin. The US Nuclear Regulatory Commission, in recognition of this problem, sponsored an experimental program at Sandia National Laboratories (SNL). A part of the program is presented in the paper.

One of SNL's primary objectives in the study was to thoroughly evaluate the correlation between analytical ultimate response prediction and experimental results. The proposal was that if both the analyses and the experiments were performed "at the same temperature", the correlation study would be as valid as if an elevated temperature had been applied.

In order to isolate the most significant parameters and to keep the study economically feasible, numerous aspects of the containment structure had to be simplified. For instance, a gas, rather than saturated steam, is used for pressurisation; minor piping penetrations are omitted; and steel liner welds are not simulated exactly.

23 Design of concrete slab bridges against flexural-shear failure

R. J. COPE
University of Liverpool, UK

SUMMARY

The clauses of BS5400 dealing with the design of reinforced concrete slab bridges against flexural-shear failure are based on tests of one-way spanning members. The critical sections in a skew slab bridge lie between a bogie of the HB design vehicle and the bearings nearest the obtuse corner. These sections are subjected to rapidly varying moment and shear force distributions and are thus subjected to conditions outside the range of the data on which the code clauses are based. The results of tests on a series of skew slab models are described. The modes of failure observed, the use of analytical methods, and the current code design strength requirements are discussed.

INTRODUCTION

Values of shear stresses at a point depend on the inclination of the reference axes to the principal stress axes. The presence of shear stresses causes inclined principal tensile stresses and thus to inclined cracks in reinforced concrete beams and slabs. After cracking, the distribution of internal forces changes considerably and cracks may be subjected to shearing forces. For design purposes, the distribution of shear stresses over depth is disregarded and the distribution of vertical shearing forces is based on the predictions of linear elastic analysis. Shear failure involves the interaction of many complicated phenomena and, to reduce the problem of designing against shear failure to manageable proportions, codes of practice provide formulae for the shear strength of sections in terms of the specifiable parameters, depth, concrete strength, and normal steel ratio. As the tests on which the code formulae are based produced shear failures of entire member cross-sections, no guidance is given on the critical breadth of slab section to be considered.

The suitability of current analytical procedures for predicting the distribution of shearing forces is assessed by using test results from both an aluminium slab and cracked reinforced concrete slabs. The concrete slabs were loaded to failure, to assess the validity of the code-based predictions of flexural-shear capacity.

ANALYSIS

Reinforced concrete bridge slabs supported on elastic bearings are highly redundant structures. Their stiffness properties are dependent on the extent of both top and soffit cracking. Although the predictions of linear elastic methods are inherently approximate, they are widely used. To provide test data with which to assess finite element and grillage solutions, a piece of aluminium was strain-gauged and tested. Details of the model slab are shown in Fig. 1.

Figure 1 Model Aluminium Slab: Geometry and Strain Gauging

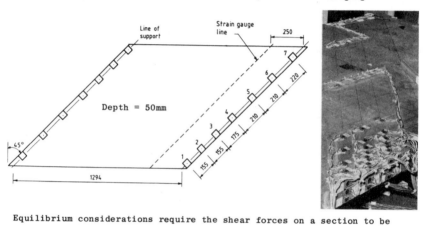

Equilibrium considerations require the shear forces on a section to be proportional to the rates of change of bending and twisting moments. The rates of change of the latter can be particularly large close to a free edge, as the shear strains fall to zero. Measurements of surface shear strains indicated that the rapid fall in twisting moments occurs over sections equal in width to about two slab depths. Analytical solutions using finely-graded meshes of thick plate and solid finite elements give similar results and indicate correspondingly high shearing forces (1). A suitable substitute grillage requires closely-spaced members parallel to the free edge, perhaps only one slab depth apart, if the high shear forces are to be predicted. The results of thin plate theory solutions can be misleading as they do not comply with the boundary conditions (1).

CONCRETE SLAB TESTS

Whilst tests on elastic models provide data for assessing the accuracy of numerical and analogue solutions, they provide no information on the non-linear effects of cracking and yielding of steel, nor on the ultimate load. To obtain such data, tests on reinforced concrete slabs are required.

Model Details

A 30°, three 45° and one 60° skew, one-fifth scale slab models were tested (2). Each model was 100mm thick and represented a single span bridge deck with right span and width of 1.88m. The models were made approximately 1m longer than required for a single test. This enabled both obtuse corners to be failed and provided a measure of repeatability. The concrete mix was designed to produce a relatively weak concrete with the same ratio of tensile to cube strength found in prototype mixes. The cement:sand: aggregate proportions were 1:3.5:3.6 and a water to cement ratio of 0.8 was used. A maximum aggregate size of 6mm was used to reduce scaling effects on aggregate interlock behaviour.

The flexural reinforcement consisted of high yield 'Torbar' of 6, 8 and 10mm diameters. The reinforcement was designed for the load combinations specified in BS5400. However, as the critical load conditions for bending and shear are similar, additional flexural reinforcement was provided outside the shear-span. This was felt to be particularly important as flexural and shear strengths do not scale proportionately. Bars reaching an edge of a slab were provided with anchorage in the form of hooks. For all but one of the 45° skew slabs, these hooks were in vertical planes, and extended over the distances between top and soffit reinforcements.

Analytical studies suggest that the spacing and stiffness of bearings have a strong influence on the distribution of reactions. The models were supported on 75mm square steel pads, which were bedded to the under-side of the concrete through 4mm thick rubber pads. The average spacing in the critical zone was about twice the slab depth. The compressive stiffness of the model bearing arrangement averaged 50kN/mm (which represents 250kN/mm for the prototype). However, the bearing stiffness was non-linear under low loads. The cast slab was placed on the bearings and they were then adjusted to ensure firm contact. Readings from the load cells at this stage were taken as the datum for reaction measurements.

30° Skew Slab

In the limited space available, it is not possible to consider in detail all of the tests performed. Attention will, therefore, be concentrated on some of the main features of the 30° skew slab tests. The slab geometry is shown in Fig.2a: which also shows the position of the HB bogie for the loading to failure. Before this test, five serviceability level loading cases were applied. These caused cracking on the soffit, in the mid-span zone, with crack spacing in the 60 - 80 mm range and with crack widths of about 0.03mm. For the test to failure, the dead loading and HA distributed loading were set to their ultimate limit state values and the displacement at the centre of the HB bogie was then progressively increased.

When the bogie load reached 130kN, a flexural crack at the front of the HB bogie developed into a shear crack as shown in Fig.3a. On the soffit, the edge of this crack ran perpendicular to the free edge and intersected the pre-existing flexural cracks, which tended to run parallel to the supported edge. At this stage, the obtuse corner bearing ceased to take further load, see Fig.4. Increasing the bogie load to 140kN caused the shear crack to develop for a short distance on the top surface. At 150kN, there was a sudden extension of this crack across the front face of the HB bogie as shown in Fig.3b. There was an immediate drop in the bogie load and the loss of stiffness caused by the shear crack prevented further loading. It can be seen from Fig.2b that the location of the failure crack coincides, approximately, with the position of soffit steel curtailment.

Graphs of the reactions along the supported edge adjacent to the HB bogie are shown in Fig.4. It can be seen that the bearings on the obtuse corner carried the most load. This behaviour was different to that observed with a rectangular slab, where the bearings fronting the HB bogie were the most heavily-loaded. Fig.4 also indicates that the rate of loading of the bearings is adequately predicted by linear theory up to a bogie load of about 50kN. This corresponds, approximately, to the design ultimate load for flexure. As the slab was designed using a lower bound method, using the Hillerborg, or Wood-Armer, equations, one would expect steel in some sections to yield at this stage. It is suspected that the offsets between the experimental and analytical curves are due to the initial non-linearities in the bearing stiffnesses.

Figure 2 30° Skew Slab: (a) geometry; (b) steel layout; (c) load-centre displacement

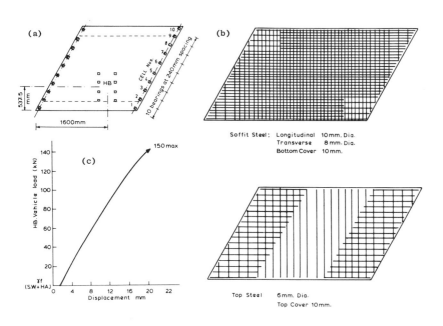

Figure 3 Cracking in the obtuse corner region

(a) at 130kN   (b) at 150kN

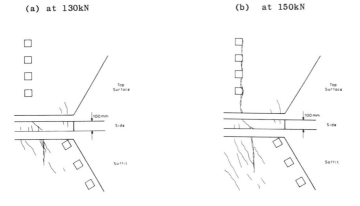

The opposite free edge of this slab was strengthened with an 'edge-beam', consisting of 3mm diameter wire stirrups at 70mm centres. The notional width of the edge beam was 100mm and 3mm corner bars were provided to assist in placing the stirrups.

The slab and loading were repositioned to facilitate a test on the opposite obtuse corner zone. Up to a bogie load of 140kN, the behaviour was similar to that observed during the first test. However, when the load was nearing 140kN, there was evidence of crushing on the top surface between the free edge and the nearest "wheel' of the HB bogie, see Fig.5. This did not affect the distribution of reactions. At 150kN, a local shear failure was observed between the bogie and the bearings. The edge of the crack ran for approximately 45mm perpendicular to the free edge on the top surface. On the soffit, it ran around the two bearings closest to the free edge, see Fig.5b. The slab was able to support increasing loads, up to 195kN, with little outward sign of worsening damage. At this load level, there was a sudden failure, with a relatively large shear displacement on a crack which ran around the four bearings closest to the obtuse corner.

Figure 4 Reactions   Figure 5 Cracking: (a) at 150kN
                                           (b) at Failure

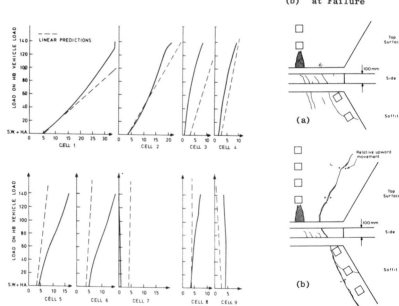

Both tests resulted in flexural shear failures. In the first test, visible shear cracking began at the free edge with the bogie load at 130kN, whereas in the second test it was initiated at 150kN. The failure load in the first test, for which there was no additional shear reinforcement, was 150kN. At this load intensity, the average shear stresses across the entire slab were $0.8N/mm^2$ on a section normal to the free edges and $0.7N/mm^2$ on a section parallel to the support line. Approximate values of shear force intensities can be obtained by considering the measured reactions. In Table 1 "shear-stresses" obtained by dividing reactions by the cross-sectional areas defined in Fig.6 are listed.

Figure 6 Trial Sections

Table 1 Average 'shear-stresses' at 140kN

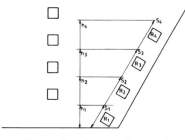

	v_{c1}	v_{c2}	v_{c3}	
Test 1	2.06	1.50	1.09	
Test 2	1.68	1.49	1.20	
	v_{c4}	v_{c5}	v_{c6}	v_{c7}
Test 1	2.45	1.77	1.32	1.05
Test 2	1.94	1.72	1.45	1.20

$v_{c1} = R_1/S_1d$; $v_{c2} = (R_1+R_2)/S_2d$; $v_{c3} = (R_1+R_2+R_3)/S_3d$
$v_{c4} = R_1/h_1d$; $v_{c5} = (R_1+R_2)/h_2d$; $v_{c6} = (R_1+R_2+R_3)/h_3d$;
$v_{c7} = (R_1+R_2+R_3+R_4)/h_4d$.

The bridge code (BS5400) suggests that the shear capacity is given by $V_c = \xi_s \gamma_m (f_{cu}\rho)^{1/3} bd$. With $\xi_s = 1.5$; $\gamma_m = 1.25$; f_{cu} = average measured cube strength; (mean $f_{cu} = 32.33N/mm^2$, standard deviation = $1.29N/mm^2$, from 12, 100mm cube tests) and $\rho = \sum_i A_{si} \cos^4 \alpha_i$ (where α_i is the inclination of the bar direction to the normal of the section under consideration), the calculated capacities are $1.21N/mm^2$ for sections normal to the free edge and $0.96 N/mm^2$ for sections parallel to the line of supports. The steel areas have been resolved in this way, as it is the bar stiffnesses that influence the crack spacing and hence the aggregate interlock and dowel actions. In performing the resolution, it has been assumed that the strains normal to the crack direction are relatively small in comparison to the normal strain across the crack. As the top surface was almost crack free, these assumptions are reasonable.

From these results it can be seen that reasonable section widths to consider in design would be the distance from the free edge to the extremity of the HB bogie, or its projection on the line of bearings. Over short section lengths close to the free edge, the code-based shear capacities were almost certainly exceeded, without detriment to the slab. The presence of the anchorage hooks on the flexural reinforcement and the nominal edge stirrups, mobilised in the second test, probably enhanced the concrete strength in the most highly-stressed region by their confining actions. The presence of 'transverse' reinforcement enabled a section of considerable length to be utilised in resisting the shear forces.

45° Skew Slabs

Both obtuse corners of three solid 45° skew slabs were tested. Slabs S1, S2 had reinforcement parallel and perpendicular to the supported edge. In S2, the anchorage hooks were located in horizontal planes. The third slab S3 had steel parallel and perpendicular to the free edges. None of these slabs was provided with stirrups.

Table 2   Critical Bogie Loads (kN)

Slab	S1	S1	S2	S2	S3	S3
First visible shear crack	80	85	70	70	90	80
Failure	107	100	80	70	110	100

The loads at which the first visible shear cracking and failure occurred are listed in Table 2. It is immediately apparent that the loads are smaller than those for the 30° slab. This is due to the greater twisting moment and shear force intensities in the shear span. It can also be seen that the load capacity after the initial shear cracking is relatively small, and almost non-existent in the slab with horizontal anchorage hooks. The initial shear cracks formed at the free edge. On the top surfaces, they ran perpendicular to the edge, but on the soffit they tended to follow the line of the bearings. At failure, the top edge of a shear crack extended to follow the line of the bearings, while on the soffit the bottom edge tended to isolate the most heavily-loaded bearings in the obtuse corner.

60° Skew Slab

The 60° skew slab failed by punching around the obtuse corner bearing as shown in Fig.7. At 35kN, an inclined crack appeared on the free edge which reached neither the top nor soffit surfaces. With further loading, it extended and by 60kN relative shear displacements were discernible across cracks on the top surface, as indicated in Fig.7. At 75kN, a complete punching perimeter formed, isolating the obtuse corner bearing.

Figure 7   (a)   Visible Cracking   (b)   at Failure
                   at 60kN

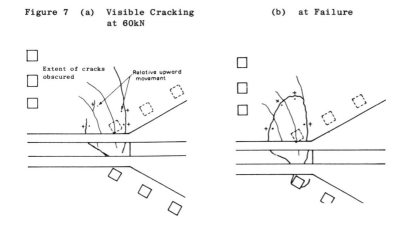

OBSERVATIONS

For a skew bridge slab, flexural-shear cracking caused by relatively concentrated loading does not result in immediate, overall structural failure. For slabs with a skew of greater than about 30°, the code design clauses are difficult to apply: the section length to be considered for flexural/torsion-shear failure is not obvious; and the use of a large punching perimeter in the vicinity of closely-spaced bearings does not seem to be appropriate. For engineers to have confidence in codes of practice, it is vital that simple methods should only be extended to new problems after experimental tests have been conducted to assess their suitability.

ACKNOWLEDGEMENTS

The author is indebted to the Science and Engineering Research Council and to the Department of Transport for financial support. Thanks are also due to Dr. P. V. Rao and Dr. K. R. Edwards for their invaluable assistance.

REFERENCES

(1) COPE, R. J., RAO, P. V., 'Shear forces in edge zones of concrete slabs', The Structural Engineer, Vol. 62A, No. 3, (March, 1984).

(2) COPE, R. J., RAO, P. V., EDWARDS, K. R., 'Shear in skew reinforced concrete slab bridges', Report to the Department of Transport (1983).

DISCUSSION

D M Porter   University College   Cardiff   UK

I was interested to hear Dr Cope describe some of the shear failures as being ductile. On a recent test that I have carried out on a thin slab, I also observed a ductile shear failure.

Section 3 Papers 17–23

Commentary by Dr J. B. MENZIES
(Assistant Director, Building Research Establishment, UK)

The papers presented in this Section on the application of model results to design illustrate a number of valuable ways of using model analysis. We can see how models of a small scale, necessitated for economic reasons, can provide insights into behaviour which may not only be helpful in design but also may have considerable educational benefit for the students, the future designers, carrying out the tests. Examples of models used for the validation and development of theoretical analyses and of models used where no good theory is available are also described. In addition, the papers include useful discussion on the relationships of cost to model scale, the limitations of models, problems of interpretation of test results and calibration against the full scale.

24 Laboratory testing of bored and cast-in-situ microconcrete piles in clay to study shaft adhesion

W. F. ANDERSON
University of Sheffield, UK
K. Y. YONG
National University of Singapore
J. A. SULAIMAN
University of Sheffield, UK

SUMMARY

Instrumented laboratory scale tests have been carried out in which an element a bored and cast-in-situ pile has been constructed simulating field procedure and after concrete curing it has been load tested to measure shaft adhesion. Results indicate that with time the horizontal effective stress approaches 'at rest' value and that shaft adhesion is governed by the residual angle of shearing resistance.

INTRODUCTION

A fundamental approach to estimating shaft adhesion on piles in clay is to consider the problem in terms of effective stresses. Chandler (1) proposed the unit shaft adhesion, τ_s, could be given in terms of the effective stress parameters, c' and ϕ',

$$\tau_s = c' + \sigma_H' \tan \phi'$$

where σ_H' is the horizontal effective stress acting on the pile and is assumed equal to $K_s . \sigma_V'$, where K_s is an earth pressure coefficient at the pile/soil interface and σ_V' is the effective overburden pressure. For normally consoli clay c' will be zero, and by assuming remoulding during pile installation in consolidated clay reduces c' to approaching zero, then $\tau_s = K_s . \sigma_V' \tan \phi'$. By assuming that the clay is normally consolidated and that the earth pressur coefficient is the 'at rest' value, Burland (2) was able to modify this equat to the form

$$\tau_s = \beta . \sigma_V' \text{ where } \beta = (1 -\sin \phi') . \tan \phi'$$

Values of β back figured from load test data for a wide range of pile and soi types were found to exhibit much less scatter than the values of the widely u total stress adhesion factor, α, thus demonstrating the potential of the effective stress approach. However, there is still considerable uncertainty about the value which should be used for the earth pressure coefficient, K_s, and whether the peak angle of shearing resistance, ϕ', governs the mobilised shaft adhesion.

The huge cost of the very long piles used for offshore oil platforms has stimulated research into the effective stress method for driven piles (e.g. 3). Particular interest has been focused on the changes in effective stress which occur during pile driving, during the period after driving as the excess pore pressures generated during driving dissipate and during pile loading.

When bored piles are installed, excess pore pressures are not set up as they are during pile driving, and remoulding of the soil close to the pile wall is much less significant. Stress relief during boring sets up negative pore water pressures in the soil around the borehole, and the high water cement ratio used for concreting bored piles means that there may be moisture migration from concrete to soil with consequent soil softening. Clayton and Milititsky (4) have reviewed some of those effects.

Compared with driven piles little work has been done on effective stress analysis of bored and cast in-situ piles. Weltman and Healy (5) and Searle (6) have reported results of some field load tests which have been reanalysed in terms of effective stress and O'Riordan (7) has reported results from a test on a single instrumented bored pile.

Problems with instrumented field studies are that they are so expensive that the test programmes have to be very limited, and load test results may be difficult to interpret because of natural soil variability. Laboratory studies can overcome these problems in that tests can be carried out with different types of soil beds carefully prepared with known stress history so as to give reproducible results. Considerable instrumentation can be used in the tests and the large number of tests which can be carried out in the laboratory means that different construction techniques can be used and their effects on the pile capacity examined.

TEST EQUIPMENT AND PROCEDURES

Since only shaft adhesion was being examined and to minimise load transfer effects along the length of the pile, a pile element was simulated as shown in Fig.1. End

Figure 1 Simulation of pile element

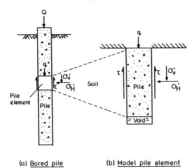

(a) Bored pile   (b) Model pile element

bearing beneath the element was eliminated by creating a void at the base. This pile element was constructed and tested by simulating field procedures using a bed of clay prepared with known stress history and maintained at realistic field stress level

To get a full understanding of the changes occurring during all construction stages total stresses in the clay bed and at the pile/soil interface were monitored using strain gauged diaphram cells. Pore water pressures were monitored with minature piezometer probes or minature pore pressure transducers at the pile/soil interface and at different radial distance from the pile elements. All instruments were connected to a data logger and scanned at appropriate intervals during a test. Details of the design, construction and calibration of the test cells and instruments have been given by Yong. The test cells were prepared by bolting the cell base to the body, which had been lightly greased to minimise friction during clay bed preparation, and the soil instruments positioned. A clay slurry prepared at twice the liquid limit was placed in the cell and consolidated in stages using water pressure as shown in Fig.1, Stage 1. When consolidation (or swelling for overconsolidated samples) was complete the clay bed had reduced in thickness to about 200mm. At this stage the test cell was inverted and a central disc removed from the end plate to expose the clay bed.

Three different types of construction technique, i.e. without casing, with casing and with bentonite, were simulated during the laboratory test programme.

The field boring process was simulated by using an auger mounted in a drilling frame attached to the end plate of the test cell, Fig. 2, Stage 2. When the hole had reached the required depth, about 120mm, the base 'void', consisting of two aluminium discs connected by a rubber membrane, was lowered to the bottom and filled with water which could be pressurised.

Because of the small volume of concrete required and the exaggeration of pile surface roughness which would have occurred if normal aggregates had been used, microconcrete (9) was used. A high workability concrete is required for place

Figure 2 Test equipment and construction sequence

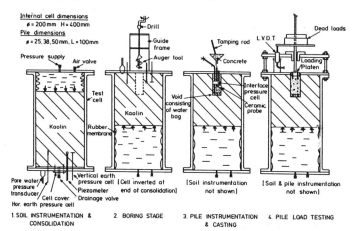

Table 1 Microconcrete mix design and properties

Water/cement ratio		0.6
Aggregate/cement ratio		2.4
Aggregate size range	2mm-1.18µm	15%
	1.18mm-600µm	35%
	600µm -300µm	30%
	300µm -150µm	20%
Slump (mm)		150
Density (Mg/m³)		2.13
7 day cube strength (N/mm²)		25.0
28 day cube strength (N/mm²)		32.5

in boreholes, so the mix given in Table 1 was used as standard. Cured microconcrete properties are also given in Table 1. Some tests were carried out at water/cement ratios of 0.5 and 0.7 to examine the effects of mix design on the moisture migration at the pile/soil interface.

The microconcrete was hand mixed and placed to about one third to one half depth of the pile using a funnel, and an interface pressure cell and pore pressure device were then put in position. The remainder of the microconcrete was then placed in the borehole, Fig. 2, Stage 3. In the field the fall of the concrete is normally considered adequate for compaction of the concrete, but because of the small fall in the laboratory the microconcrete was tamped gently to promote intimate contact with the soil and to prevent arching around the instruments. A loading platen was placed on top of the pile element and the concrete pressurised to simulate the pressure of the fresh concrete at depth. Load testing of the pile element, as shown in Fig. 2, Stage 4, was usually carried out after 7 days, although a limited number of tests were carried out after longer cure periods, up to a maximum of about 10 months. The load was applied by dead weights put on a loading platform which transmitted the force through a piston to the load cap seated on the concrete. The settlement was monitored by two displacement transducers mounted diametrically opposite to ea other. A maintained load technique was used, with the magnitude of increment being reduced as settlements became more significant. Each increment was appl until the rate of settlement had reduced to less than 0.01mm in five minutes, giving load tests of 4 to 6 hours duration.

TYPICAL TEST RESULTS

To give an indication of the type of information which is available from these tests some typical results are given. Detailed results and interpretation wil be found elsewhere (10).

Figure 3   Variation of horizontal stresses with time

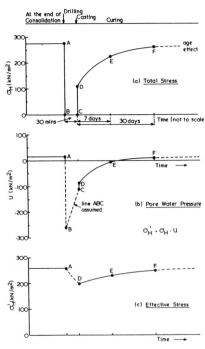

Earth pressure changes

The horizontal and interface earth pressu cells allow the total horizontal stress a all stages of the test to be known. Pore water pressure measured in the soil mass at the pile soil interface can be monito for most of the test, although the large negative pore water pressures created du stress relief on boring have to be assume By subtracting the pore water pressures the total stresses, the effective stress be found. Fig. 3 shows the stress chang which occur in a typical test and indica that although there is a reduction in th radial effective stress during construct with time the effective stress increases again and approaches the 'at rest' value In tests on both normally and overconsol dated clays, if the delay between boring concreting did not exceed 30 minutes, th some 85-90% of the 'at rest' pressure wa reached within 7 days. Indirect measure in long term tests indicated full regain earth pressure at about 70 days.

Mobilised angle of shearing resistance

Table 2 gives the analysis of the results of one series of tests in which uncased holes were used during pile construction in kaolin beds prepared with different stress histories. By assuming that the effective cohesion, c', is zero, the value of the mobilised angle of shearing resistance, δ, may be calculated from the measured horizontal effective stress and the shaft adhesion.

Also included in Table 2 are values of the residual angle of shearing resistance for the kaolin found from ring shear tests. It can be seen that the mobilised angles of shearing resistance are approaching the residual values. None of the pile load-settlement curves indicated a peak behaviour and no excess pore pressures were measured during loading indicating drained conditions in the thin annular shear zone.

The results suggest that even with bored piles where the clay fabric disturbance is very much less than with driven piles, considerable reorientation of the clay particles occurs during boring. Chandler and Martins (11) in model tests using resin cast in clay holes prepared with minimum disturbance have found that a very thin (0.5mm) continuous shear zone is formed adjacent to the pile face. These results would indicate that boring may reorientate the particles in this narrow zone prior to pile construction and loading, so that only the residual shearing resistance has been mobilised during load testing.

Table 2 Comparison of mobilised angle of shearing resistance with residual angles of shearing resistance from ring shear (ϕ'_{peak} = 23°)

Overconsolidation Ratio	Measured shaft adhesion kN/m²	Measured earth pressure coefficient, K	Mobilised angle of shearing resistance	Residual angle of shearing resistance
1	19.1	0.53	14.4°	13.0°
2	23.8	0.71	13.5°	12.3°
3	28.0	0.89	12.7°	12.1°
4	31.6	1.03	12.4°	11.4°
5	34.9	1.16	12.1°	11.3°
6	38.5	1.31	11.9°	11.1°
7	42.1	1.45	11.7°	11.0°

Moisture migration studies

Moisture migration studies have shown that with a water/cement ratio of 0.6 there may be a slight increase in the water content of the soil around the pile element after seven days. This is in accordance with the earth pressure coefficient values being slightly below 'at rest' values i.e. some stress relief and soil softening has occurred. With further curing time the soil water content decreases

Table 3 Variables whose effects were examined

Clay type
Overburden pressure
Overconsolidation ratio
Use of dry boreholes
Use of cased boreholes
Use of bentonite
Delay between boring and concreting
Cure time before testing
Water/cement ratio
Pile diameter

indicating an increase in effective horizontal stress, and stabilises at a value slightly below the original moist content value prior to construction. T would imply that the cured concrete may acting as a drain for soil water.

Tests carried out with a water/cement ra of 0.7 gave an increase over the 0.6 te of about 2% in soil water content close to the pile after seven days. The tes carried out with a water/cement ratio of 0.5 gave a decrease of similar magnit below the 0.6 test values.

DISCUSSION AND CONCLUSIONS

The approach to modelling used in this study is of the Category II type sugges by James (12). In this category the model may be considered as a small protot structure, and its behaviour can be compared with that predicted by some metho of analysis. As James has pointed out, the results obtained are not necessari of immediate use in the design of a complex prototype, but they are of great value in establishing certain design principles. In this study this approach allowed the effective stress method to be examined for the wide range of variab listed in Table 3. So far, in excess of 100 tests have been carried out. Fi testing on this scale would be impossible, and it would be unlikely that the s control could be kept over the variables. The success rates of the instrument tion would also be lower with field testing.

The results have highlighted a number of trends. Of particular importance are the fact that the 'at rest' horizontal effective stress appears to be substantia restored within a relatively short period of time and that boring may cause sufficient remoulding of the soil close to the pile face to mean that only the residual shearing resistance may be mobilised.

The limitations of laboratory scale testing should not be overlooked. True modelling can only be achieved by centrifuge testing, but like field tests the are quite expensive. In these particular tests an attempt was made to use realistic field stress levels, so the piles should be considered as minature, rather than model. Because of this one series of tests was carried out using different diameters of pile element. The unit shaft resistances at any settlem varied by no more than 1½%, but settlements to failure increased with increasing diameter so that failure loads varied by ±6% of the mean value. This would indicate that scale effects are relatively small and confirms Meyerhof's view that shaft adhesion is not unduly affected by pile size (13). Elsewhere (10) field test data has been examined in the light of the trends found in the laboratory studies. However most published field data was recorded for total stress analysis and gross assumptions have to be made when analysing the data in terms of effective stress. Also progressive failure leading to load transfer effects complicates the back analysis. In order to confirm the use of the effective stress method for bored piles and allow designers to use it with confidence, further laboratory studies are required, but more importantly the evidence of all these studies must be checked in a carefully planned programme of fully instrumented field load tests on pile elements.

REFERENCES

(1) CHANDLER, R J, 'The shaft friction of piles in cohesive soils in terms of effective stress', Civil Engineering and Public Works Review, Vol. 63, pp. 48-51 (1968)

(2) BURLAND, J B, 'Shaft friction of piles in clay: a simple fundamental approach', Ground Engineering, Vol. 6, No. 3, pp. 30-42 (1973)

(3) KIRBY, R C, ESRIG, M I and MURPHY, B S, 'General effective stress method for piles in clay. Part I - Theory', Proc. Am. Soc. Civil Engrs Conf on Geotechnical Practice in Offshore Engineering, pp. 457-498 (1983)

(4) CLAYTON, C R I, and MILITITSKY, J, 'Installation effects and the performance of bored piles in stiff clay', Ground Engineering, Vol. 16, No. 2, pp. 17-22 (1983)

(5) WELTMAN, A J, and HEALY, P R, 'Piling in boulder clay and other glacial tills', CIRIA Report PG5 (1978)

(6) SEARLE, I W, 'The design of bored piles in overconsolidated clays using effective stresses' Proc. Conf. on Recent Developments in the Design and Construction of Piles, Instn Civil Engrs, London, pp. 355-364 (1980)

(7) O'RIORDAN, N J, 'The mobilisation of shaft adhesion down a bored, cast-in-situ pile in the Woolwich and Reading beds', Ground Engineering, Vol. 15, No.3, pp. 12-26 (1982)

(8) YONG, K Y, 'A laboratory study of the shaft resistance of bored piles', Ph.D. Thesis, University of Sheffield (1979)

(9) ALDRIDGE, W W, and BREEN, J E, 'Useful techniques in direct modelling of reinforced structures', Am. Concrete Inst., Rep. No. 24 (1968)

(10) ANDERSON, W F, YONG, K Y, and SULAIMAN, J A, 'Shaft adehsion on bored and cast-in-situ piles in terms of effective stress' In press (1985)

(11) CHANDLER, R J, and MARTINS, J P, 'An experimental study of skin friction around piles in clay', Geotechnique, Vol. 32, No. 2, pp. 119-132 (1982)

(12) JAMES, R G, 'Some aspects of soil mechanics model testing', Proc. Roscoe Memorial Symposium on Stress-strain Behaviour of Soils, Cambridge, pp. 417-440 (1972)

(13) MEYERHOF, G G, 'Scale effects of ultimate pile capacity', Proc. Am.Soc. Civil Engrs, J. Geot Engng, Vol. 109, GT6, pp. 797-806 (1983)

DISCUSSION

D P Abrams   University of Colorado   USA

Numerous papers in this volume report applications where reduced-scale models have been used for engineering research. Models were fabricated with an attempt to represent behaviour of silos, pile caps, blast shelters, and components of buildings. With the exception of a few papers, the accuracy of the model to represent prototype behaviour for the particular loading has not been questioned. Apart from verifying numerical models, test results from a reduced-scale model must be viewed in terms of the accuracy of the model and the sensitivity of results to this accuracy. If this is not done, the usefulness of the study is limited to model research rather than the construction of engineered structures.

A large amount of experimental research has been done using large-scale specimens, particularly in the United States and Japan for earthquake engineering. Scaling relations could be established using these tests as benchmarks by fabricating direct models at a reduced scale. It would appear that knowledge of the precise factor at which a particular phenomena may be modelled would be essential to the determination of experimentation costs, or assessment of capabilities of an existing facility. Perhaps a future symposium that would gather individuals with interests in large-scale testing and modelling may stimulate new interests for researchers from both camps.

25 Model analysis of grouted connections for use in construction and repair of offshore structures

L. F. BOSWELL and C. D'MELLO
Department of Civil Engineering, City University, London, UK

SUMMARY

The grouted connection between the piles and sleeves of a fixed offshore platform is the major structural connection between the foundation and platform. Split grouted connections may also be used to repair structural members. This paper discusses the way in which the ultimate load behaviour of protype connections may be obtained from composite models. The results of some model tests are presented and discussed.

INTRODUCTION

Grouted connections have been used as the main structural connection between the piles and the jacket of fixed offshore platforms. Increasingly, split sleeve connections are being used to strengthen or repair structural members. In this latter case the sleeves are bolted either side of the member and the annulus between the member and the sleeve is filled with cement grout. Figure 1 shows the important dimensions of a connection.

The structural testing of grouted connections is essential in establishing their ultimate load characteristics. In general it is not practical to examine the behaviour of full scale specimens, their dimensions being too great for normal laboratory considerations. Model tests are, therefore used to provide essential design data. In fact, the results of design recommendations[1] have been obtained from model testing.

DESCRIPTION OF MODELS

The specimens which have been examined in the laboratories at The City University have been conducted at a 1/4 and 1/5th scale and have been constructed of the same materials as the prototypes. Oilwell B cement with a plasticiser has been used for the cement grout and the steel members have been produced from standard or rolled tubulars.

The strength of a connection depends upon the stiffness of the grouted tubulars, the contact grout area in shear, the bond between the grout and steel interface and the height of the shear connectors, if these have been used.

The models were geometrically similar, the D/t and h/s dimensions being reduced by the scale factor. Since the same materials were used for both prototype and models, the previously mentioned factors affecting strength were also scaled proportionally. In particular, the dimensions of the models were such that the average interface grout-steel shear bond stress was the same as the prototype. The grout mix design used a 0.36 W/C ratio with a 1.5% super plasticiser by weight.

The model specimens were immersed in water and cured at 8°c until they were tested. These conditions were identical as far as possible to those of the prototype.

Certain difficulties associated with the construction of scale models are unavoidable and produce scale effects. For grouted connections these were the possibility of local buckling of the tubulars and the method of load application. The fact that the connections were not tested underwater was considered unimportant. In order to avoid local buckling, longitudinal stiffeners were positioned around the circumference of the outer tubulars. In most cases the loads applied to prototype connections are transferred along the length of the connections. For models, the loads were applied at the ends. The arrangement of stiffeners and the method of load application may be seen in Figure 2. The additional unwanted stiffness provided by the longitudinal stiffeners was taken into account to calculate an effective stiffness of the tubulars.

It was considered that the dimensions of the scaled specimens were such that effects due to particle sizes in the grout were insignificant. In anycase it is known that the interface between the grout and steel has a considerable effect on the mode of failure particularly for the case of plain pipes without shear connectors. It is only during the latter stages of failure that crushing takes place within the grout and then only in the zones beneath the shear connectors.

DESCRIPTION OF THE MODEL TESTS

A series of tests has been conducted in which the load has been applied in an axial direction for all specimens. Table 1 provides the important dimensions of the models. These dimensions have been especially selected to augment a previous research programme[2] and the behaviour of both plain pipe specimens and specimens with shear connectors has been investigated.

The following tests have been conducted, (i) Four compression tests on specimens with extreme geometries, (ii) Three tension tests on plain pipes and (iii) Three tension tests on connections with shear connectors.

The objective of the first series of tests was to examine the applicability of current design formula to a wider range of connection geometries than had been examined previously. In particular, the behaviour of extremely stiff and extremely flexible connections has been examined. Also the effect upon performance of a large grout annulus and large shear connector height (welded bead) has been investigated.

The application of grouted connections in tension or guyed platforms requires the joint to be subjected to tensile forces. In the original use of connections as a joint between the pile and sleeves of a fixed platform, the joint is subjected to compressive forces. Since all previous results have been related to the compression mode, a comparison has been undertaken in which plain pipe joints (series ii) and joints with shear connectors (series iii) have been subjected to tensile forces and the results of their behaviour compared with geometrically similar compression tests. The general arrangement of the system of loading specimens can be seen in Figure 2. The loads were applied in increments which decreased in magnitude as failure was initiated. The specimens were re-loaded after the initial failure and the ultimate failure was assumed to be the maximum load achieved.

Table 1 Dimensions of test specimens

Specimen	L/D	Pile		Sleeve		Grout		D/t Ratios			Shear connectors		
		OD	WT	OD	WT	OD	WT	Pile	Sleeve	Grout	Height	Spacing	h/s
A1	2	508	25	568.76	4.98	558.8	25.4	20.32	114.21	22	2.03	169.33	0.012
A2	2	508	16	568.76	4.98	558.8	25.4	31.75	114.21	22	2.03	169.33	0.012
A3	2	508	12.5	568.76	4.98	558.8	25.4	40.64	114.21	22	2.03	169.33	0.012
B1	2	508	25	568.76	4.98	558.8	25.4	20.32	114.21	22	-	-	-
B2	2	508	16	568.76	4.98	558.8	25.4	31.75	114.21	22	-	-	-
B3	2	508	12.5	568.76	4.98	558.8	25.4	40.64	114.21	22	-	-	-
C1	2	508	16	689.4	6.03	677.33	84.67	31.75	114.33	8.0	2.03	169.32	0.012
D1	2	508	16	568.76	4.98	558.8	25.4	31.75	114.2	22	7.11	169.32	0.042
E1	2	508	30	598.8	20.0	558.76	25.4	16.9	29.94	22	2.03	169.32	0.012
F1	2	508	9.5	598.8	20.0	558.76	25.4	53.5	29.94	22	2.03	169.32	0.012

MODEL INSTRUMENTATION

Linear voltage displacement transducers and direct reading dial gauges were used as a measure of the relative movement between the pile and the sleeve of a specimen.

The strains were measured by electrical resistance gauges and all specimens were examined extensively. Gauges were positioned to measure both the circumferential and longitudinal strains. The specimens subjected to tension had gauges fixed to both pile and sleeve whilst for compression specimens had gauges fixed only on the sleeve.

A compulog IV automatic data logging system was used to record the experimental information throughout a test. The load which was recorded by measuring the pressure in each hydraulic jack was compared with a direct load cell measurement.

RESULTS

A summary of all the test results is given in Table 2, together with the grout strengths and specimen geometric ratios. The grout strength was determined from three inch cubes which were tested at the same time as the specimens. For each of the tests, load-slip graphs were plotted and an example is shown in Figure 3. In all cases the specimen was reloaded after initial failure and again after subsequent failures until a maximum load was achieved or until substantial slip had occured.

A bond stress, which is defined as the failure load divided by the bond area, is calculated at ultimate failure. Since the grout strength varies with age it is necessary to normalise the bond stress with respect to the cube strength. A non-dimensional term has been defined in previous tests [2] and is given by:-

$$F_{bu} = \frac{f_{bu}}{f_{bu}^{API}} \left(\frac{50}{f_{cu}}\right)^{0.5}$$

where f_{bu}^{API} = API design bond stress

f_{cu} = cube strength

f_{bu} = ultimate bond stress

The results are presented in more detail in what follows.

Table 2   Test Results

Specimen	D/t			D_p/S	h/D_p	h/s	L/D	Age (Days)	f_{bu} N/mm²	f_{cu} N/mm²	F_{BU}
	Pile	Sleeve	Grout								
A1*	20.32	114.21	22	3.0	0.004	0.012	2	8	2.23	71.3	1.69
A2*	31.75	114.21	22	3.0	0.004	0.012	2	11	2.01	66.1	1.58
A3*	40.64	114.21	22	3.0	0.004	0.012	2	9	2.01	70.6	1.53
B1*	20.32	114.21	22	-	-	-	2	11	0.51	61.1	0.40
B2*	31.75	114.21	22	-	-	-	2	9	0.27	62.3	0.22
B3*	40.64	114.21	22	-	-	-	2	10	0.47	67.0	0.37
C1	31.75	114.33	8.0	3.0	0.004	0.012	2	4	3.03	41.5	3.00
D1	31.75	114.21	22	3.0	0.0014	0.042	2	2	2.06	9.52	4.27
E1	16.9	29.94	22	3.0	0.004	0.012	2	3	3.38	25.4	4.30
F1	53.5	29.94	22	3.0	0.004	0.012	2	7	2.79	54.6	2.42

TENSION TESTS

(i)   PLAIN PIPE SPECIMENS

All the specimens had similar load slip curves (Figure 3) with the maximum load occuring at the initial failure. This was to be expected as these connections rely upon an adhesive bond between the steel and grout surfaces.

The associated bond stresses were relatively low and showed some amount of scatter. This was in part due to grout shrinkage which was observed as radial cracks.

(ii) WELD BEADED SPECIMENS

The load-slip curves for these specimens were similar to that shown in Figure 4. A comparison of these results with the results obtained from the plain pipe connectors demonstrates the great difference in failure loads. A comparison using the parameter F_{bu} is also made with identical tests on specimens in compression and these results are shown in Table 3. It can be seen from this table that the bond stress occurring in the tension specimens is approximately 15% less than that in the compression specimens.

Table 3   A Comparison between Tension and Compression Results

Tension Specimens F_{bu}	Compression Specimens F_{bu}	% Difference
1.69 (A1)	2.00	15.4%
1.58 (A2)	1.85	14.8%
1.53 (A3)	1.81	15.4%

COMPRESSION TESTS ON EXTREME GEOMETRY SPECIMENS

These tests were carried out to extend the range of applicability of current design rules. Only limited conclusions can be drawn from the results of single tests but when included with other test data, more general conclusions can be obtained.

(i) LARGE GROUT ANNULUS

The strength of this specimen was significantly higher than the design strength. This was consistant with the trend that a large grout annulus gives rise to higher ultimate load. This is probably due to the mechanism of failure in the grout in which diagonal cracks have to be formed before slip takes place.

(ii) LARGE SHEAR CONNECTOR HEIGHT

Previous tests[2] have shown that there is a linear relationship between shear connector height and the bond stress. This test on a specimen with large shear connector height extends the range for this relationship. The test shows that there could be levelling off of the stress at higher shear connector heights.

(iii) STIFF AND FLEXIBLE PILES

There is an increase in strength of the connection with increasing pile stiffness. The two tests conducted on the extreme pile stiffness support this conclusion although in the case of the very stiff pile this increase is small.

STRAIN RESULTS

It is not possible to present all the results recorded and only a typical example is shown. In the tension tests both the pile and the sleeve were gauged at three positions around the circumference and the readings were averaged at each height interval. For the plain pipe tests gauges were placed at regular height intervals. For the weld beaded specimens additional gauges were placed just above and below the shear connectors.

The results of a plain pipe tension specimen are shown in Figure 5 and 6 while those for an identical weld beaded specimen are shown in Figures 7 and 8. The difference in strain magnitudes reflect the significant difference in ultimate loads. The changes in strain profile were associated with crushing of the grout. Large strains were recorded around the weld beads which, at higher loads, were about two to three times the strains midway between shear connectors; the shear connectors carrying equal amount of load.

CONCLUSION

A series of tests has been conducted on scale models of grouted connections. The results of the experiments provide information on the ultimate load behaviour of prototype connections. Results of similar model tests have provided data for the design of prototype connections.

Although the procedure of scaling provided certain experimental difficulties these have been overcome and the modelling technique is a feasible and economic method of investigating prototype behaviour.

REFERENCES

1. Department of Energy. 'Report on the Working Party on the strength of grouted pile/sleeve connections for offshore structures'. Offshore Technology paper OTP11, September 1982.

2. Wimpey Laboratories Ltd. 'Department of Energy. The strength of grouted pile/sleeve connections - Phase II - Final Report - static tests'. Report No. ST22/80, April 1980.

Fig. 1   Geometry of a pile/sleeve grouted connection

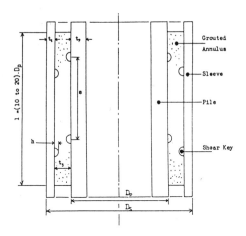

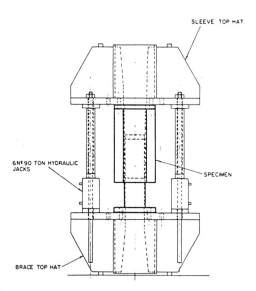

Fig. 2   General Test Arrangement

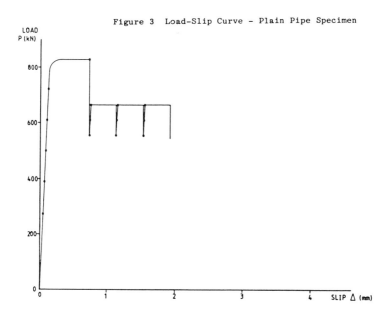

Figure 3 Load-Slip Curve - Plain Pipe Specimen

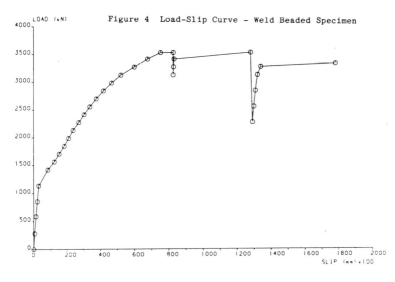

Figure 4 Load-Slip Curve - Weld Beaded Specimen

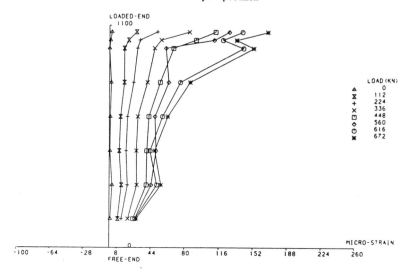

Figure 5 Sleeve Longitudinal Strain - Plain Pipe Specimen

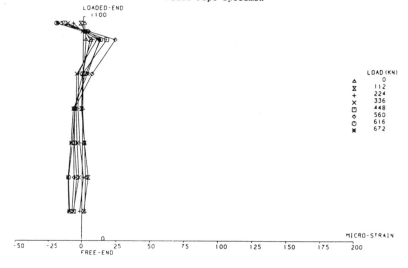

Figure 6 Sleeve circumferential Strain - Plain Pipe Specimen

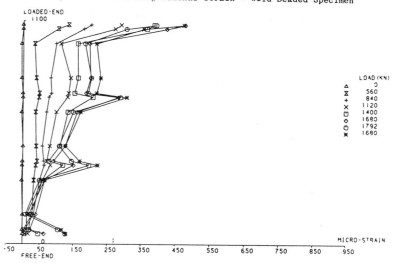

Figure 7 Sleeve Longitudinal Strain - Weld Beaded Specimen

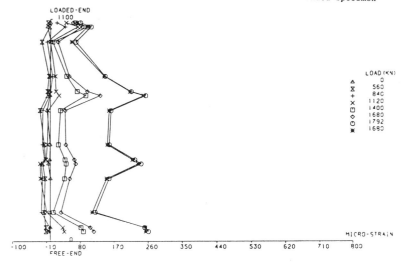

Figure 8 Sleeve Circumferential Strain - Weld Beaded Specimen

26 Reinforced concrete arch- and box-type structures under severe dynamic loads

T. KRAUTHAMMER
University of Minnesota, Minneapolis, USA

SUMMARY

Structural performance under severe dynamic loads is an important issue that has been studied extensively by many investigators. When the behavior of shallow-buried reinforced concrete arch- and box-type structures under the effects of blast loads is of interest the problem could become quite difficult. Naturally, it would be impossible to employ closed-form solutions for such studies, and one would have to combine experimental and numerical methods in order to obtain adequate information on the subject. Researchers need to consider the effects of loading conditions, material properties, soil-structure interaction, geometry and structural detailing, loading rates, structural mechanism, and others on the behavior of the system under consideration. This paper concentrates on two activities which were part of the research program, the first is the experimental studies that included the design, pretest prediction, testing of scaled structural systems, and post-test evaluation. The second consisted of developing

simple numerical techniques that are based on description of structural
resisting mechanisms for fast and accurate evaluation of system behavior.
Other activities included advanced analyses based on finite difference and
finite element techniques, but they will not be discussed here. The purpose
of this paper is to discuss and demonstrate the importance of understanding
the relationships between structural behavior and the application of model
analysis for system design. It will be shown that employing accurate
behavioral models for the analysis of reinforced concrete structures under
complex loading conditions can provide reliable information on the
structural performance.

BACKGROUND

Flat-roof structures have been studied both experimentally and through
numerical evaluations in order to understand their behavior under severe
blast loading conditions, as discussed by Kiger and Getchell (5),
Krauthammer (6), and Murtha and Holland (7). It was noticed experimentally
(5) that the roofs exhibited two types of behavior as follows: one group of
structures had a flexural response while another group was influenced
primarily by shear. At that point it became extremely important to
understand both the flexural and shear modes of behavior, and to be able to
predict that performance based on rational models. The structures were
analyzed by finite element and finite difference techniques where it was
possible to reproduce the observed behavior, nevertheless, such analyses
were expensive and time consuming. A simplified approach was proposed based
on introducing structural resistance functions into single-degree-of-free-
dom (SDOF) techniques, and including modified failure criteria for
separating flexural cases from those controlled by shear (6). That approach
seemed to produce behavior predictions that were within ten percent from

the experimental data. The quality of the numerical results indicated that it would be possible to improve the results by using more accurate descriptions of structural mechanisms and soil-structure interaction effects, as described by Bazeos and Krauthammer (2), and Holmquist and Krauthammer (4). Indeed, including explicit models for dynamic shear and thrust effects on reinforced concrete slabs has improved the numerical results to be within five percent from the experimental values.

Arch-type structures have been studied under similar loading conditions, as described by Betz et al. (3), Parsons and Rinehart (9), and Smith (10). The technical effort consisted of experimental and numerical studies of scaled test articles, and evaluation of such results in light of existing theories. The quality of numerical results based on the SDOF approach, as briefly discussed above, was a major factor in trying to apply the method for arch-type structural analysis. The report by Auld et al. (1) may illustrate one such attempt, but unfortunately the proposed models for structural behavior did not provide the same degree of accuracy as was achieved for box-type structures.

The structural systems under consideration were scaled reinforced concrete test articles that were buried under a shallow cover of soil backfill (0.6m for the systems that will be discussed here), and loaded by a pressure pulse that was applied to the soil surface. In all these cases the load was generated by a High Explosive Simulation Technique (HEST), and the peak pressures were in the range from about 10 MPa up to about 193 MPa. Test data included strains, accelarations, peak relative displacements, soil stresses, interface pressures, blast pressures, and careful examination of the test articles before and after the tests. The SDOF approach was based on employing explicit load deflection relationships for the structural elements under consideration (i.e., slabs for the box-type structures, and arches for the arch-type structures, as well as the various connections in those structures), and also failure criteria that would allow the model analysis to accurately simulate actual behavior, as presented next.

STRUCTURAL MECHANISMS

The SDOF approach as applied for pretest predictions was described by Krauthammer (6), where the resistance function for reinforced concrete slabs, and a discussion of failure criteria have been presented. As mentioned in Reference (6), one can find an excellent discussion on the topic of load-deflection relationships for reinforced concrete slabs in the book by Park and Gamble (8). The current approach is a modified version of the method that was provided in Reference (6), and includes the following developments.

Externally Applied Thrust

The peak structural resistance in the compression membrane mode can be enhanced if external in-plane compressive forces are applied to the slab, as discussed in Reference (6). For the box-type structures under consideration those forces vary with time since they are generated by the horizontal component of the pressure pulse that propagates vertically through the soil. In the present procedure the vertical pressure pulse in the soil is traced at each time step, the load on the vertical wall is computed, and then one computes the reactions in the roof and floor slabs that correspond to the applied load. These reactions are the externally applied compressive forces that enhance the peak structural resistance.

Dynamic Shear

The original shear resistance function, as proposed by Hawkins, and summarized in the report by Murtha and Holland (7) can be described as follows. Consider the straight-line segmental resistance function, as illustrated in Figure 1, and the description of these segments which are presented next.

Figure 1   Shear Stress vs. Shear Slip Model

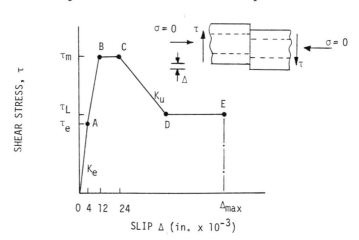

1. **From 0 to A:** the response is elastic, and the slope, K_e, of the curve defined by the shear resistance, τ_e, for a slip of 4×10^{-3} inch. That resistance is given by the expression

$$\tau_e = 165 + 0.157 f'_c \tag{1}$$

where both τ_e and f'_c are in psi. The initial response should be taken as elastic to not greater than $\tau_m/2$.

2. **From A to B:** the slope of the curve decreases continuously with increasing displacements until a maximum strength, τ_m, is reached at a slip of 12×10^{-3} inch. The maximum strength, τ_m, is given by the expression

$$\tau_m = 8\sqrt{f'_c} + 0.8\, \rho_{vt}\, f_y \leq 0.35\, f'_c \qquad (2)$$

where τ_m, f'_c, and f_y are in psi, ρ_{vt} = ratio of total reinforcement area to area of plane which it crosses; and f_y = yield strength of reinforcement crossing plane.

3. **From B to C:** the shear capacity remains constant with increasing slips. C corresponds to a slip of 24×10^{-3} inch.

4. **From C to D:** the slope of the curve is negative, and constant, and independent of the amount of reinforcement crossing the shear plane. The slope is given by the expression.

$$K_u = 2{,}000 + 0.75\, f'_c \quad (\text{psi/in.}) \qquad (3)$$

5. **From D to E:** the capacity remains essentially constant until failure occurs at a slip of Δ_{max}. For a well-anchored bar the slip for failure is given by the expression

$$\Delta_{max} = C\left(\frac{e^x - 1}{120}\right) \quad (\text{in.}) \qquad (4)$$

where

$$x = \frac{900}{2.86\sqrt{f'_c/d_c}}\,; \quad C = 2.0\,; \quad d_b = \text{bar diameter (in.)} \qquad (5)$$

and the limiting shear capacity, τ_L, is given by

$$\tau_L = \frac{0.85 \, A_{sb} \, f'_s}{A_c} \quad (\text{psi}) \tag{6}$$

where: A_{sb} = area of bottom reinforcement; f'_s = tensile strength of bottom reinforcement; and A_c = cross-sectional area.

That basic model was modified (4) to include the influence of loading rate and in-plane forces on the shear strength of reinforced concrete slabs. One can express such contributions by enhancing τ_m in Equation (2) by a factor K, and it was found that a good agreement with experimental data was obtained for K = 1.4.

Arch Resistance

The resistance functions for reinforced concrete arch-type structures has been derived similarly to that for box-type structures (6). The general form of the parabolic part (i.e., from zero load and deformation and up to peak structural resistance) is the following:

$$R = C_1 \Delta^2 + C_2 \Delta \tag{7}$$

R = Structural Resistance; Δ = Crown deflection; C_1, C_2 = Constants that depend on material properties, detailing, reinforcement ratio, geometry, and rate effects. For the structures under consideration C_1 was between 400 to 420 (psi/in^2) while C_2 was between 1300 to 1450 (psi/in).

The peak structural resistance was obtained at crown deflections that were about 0.25 of the arch thickness, and beyond that value up to a crown deflection of about 1.0 arch thickness the resistance was kept constant.

When the crown delfections exceeds the arch thickness a tensile membrane enhancement could be possible, but such behavior was not obtained in the current tests.

EXPERIMENTAL - NUMERICAL COMPARISON

Experimental and numerical results for box-type structures are presented in Table 1, and the comparison demonstrates the effectiveness of employing SDOF methods for the analysis of experimental models. Similar quality results were obtained from the analyses of arch-type structures.

Table 1 Summary of Results

Event	Experiment Structural Behavior or Failure Mode	Measured Permanent Deflection Δ (mm)	Average Peak Pressure (MPa)	Computed Permanent Deflection Δ1 (mm)	Δ1/Δ	Analytical Shear Slip Attained (mm)	Failure Shear Slip (mm)	Time of Failure (m sec)
No. 1	Flexure	12.7	9.0	16.2	1.28	0.14	5.6	—
No. 2	Shear	Failed	36.2	Failed	1.0	5.0	5.0	1.02
No. 3	Flexure	152.4	18.3	152.4	1.0	0.16	4.8	—
No. 4	Flexure	304.8	20.0	309.9	0.98	0.4	6.0	—
No. 5	Flexure	78.4	79.0	81.3	1.03	20.3	29.7	—
No. 6	Shear	Failed	80.2	Failed	1.0	5.8	5.8	1.06
LB	Flexure	185.4	8.0	182.9	0.99	0.33	3.8	—
				Average:	1.04			

298

REFERENCES

(1) AULD, H E, DASS, W C, and MERKEL, D H, Development of Improved SDOF Analysis Procedures for Buried Arch Structures, Air Force Weapons Laboratory, Final Report No. AFWL-TR-83-39, August 1983 (limited distribution).

(2) BAZEOS, N, and KRAUTHAMMER, T, An Improved Numerical Approach for the Analysis of Reinforced Concrete Slabs Under the Effects of Dynamic Loads, University of Minnesota, Department of Civil and Mineral Engineering, Structural Engineering Report No. 83-01, March 1984.

(3) BETZ, J F, et al., Kachina Test Series: Eagle Dancer, Air Force Weapons Laboratory, Final Report No. AFWL-TR-83-35, Vol. I-II, July 1983 (limited distribution).

(4) HOLMQUIST, T J, and KRAUTHAMMER, T, A Modified Method for the Evaluation of Direct Shear Capacity in Reinforced Concrete Slabs Under the Effects of Dynamic Loads, University of Minnesota, Department of Civil and Mineral Engineering, Structural Engineering Report No. 83-02, May 1984.

(5) KIGER, S A, and GETCHELL, J V, Vulnerability of Shallow-Buried Flat Roof Structures, U.S. Army Engineer Waterways Experiment Station, Technical Report SL-80-7, five parts, September 1980 through February 1982 (limited distribution).

(6) KRAUTHAMMER, T, Shallow Buried RC Box-Type Structures, ASCE Journal of Structural Engineering, Vol. 110, No. 3, March 1984, pp. 637-651.

(7) MUKTHA, R N, and HOLLAND, T J, Analysis of WES FY82 Dynamic Shear Test Structures, Naval Civil Engineering Laboratory, Technical Memorandum No. 51-83-02, December 1982.

(8) PARK, R, and GAMBLE, W L, Reinforced Concrete Slabs, Wiley-Interscience, 1981.

(9) PARSONS, R, and RINEHART, E, Kachina Test Series: Butterfly Maiden, Air Force Weapons Laboratory, Final Report No. AFWL-TR-82-132, Vol. I-II, June 1983 (limited distribution).

(10) SMITH, J L, et al., Kachina Test Series: Dynamic Arch Test-3 Pretest Report, Air Force Weapons Laboratory, Final Report No. AFWL-TR-83-56, September 1983 (limited distribution).

27 Small-scale model test of a frame-shear wall structure

HELMUT KRAWINKLER and BENJAMIN WALLACE
Stanford University, California, USA

SUMMARY

This paper discusses the results and implications of a series of small- scale model tests performed as part of a U.S.-Japan cooperative research program. An assessment is made of the accuracy with which prototype behavior is reproduced in small-scale models and practical conclusions are presented that are relevant to the seismic design of frame - shear wall structures.

INTRODUCTION

In a recently completed cooperative research project, several U.S. and Japanese organizations performed a series of coordinated experiments on a reinforced concrete structure, structural assemblies and components. The seven story frame - shear wall structure shown in Fig.1 was selected as the prototype for the study. The objectives of the study were to assess various experimental techniques and to obtain detailed information on the seismic behavior of reinforced concrete structures for implementation in design and analytical modeling.

A full-scale test of the complete structure, performed at the Building Research Institute in Tsukuba, Japan, was used as the basis for correlating experimental results from the various testing programs. These testing programs included full-scale and reduced-scale (1:2, 1:3, 1:5, 1:10 and 1:12.5 scale) tests of components, assemblies and complete structures, utilizing quasi-static or pseudo-dynamic loading arrangements or shake tables as excitation sources. Detailed results of these studies are documented in many research reports, a series of papers published in the

Proceedings of the 8th World Conference on Earthquake Engineering, and a soon to be published Special Publication of the American Concrete Institute.

As part of this cooperative research effort, several 1:12.5 scale models were tested at Stanford University. The model specimens included beam- column assemblies of the type shown in Fig.2, an isolated shear wall (see center portion of Fig.1) and a shear wall - frame unit that represents the center unit of the prototype structure with a 4 m wide slab on each side. The specimens were true replica models insofar that geometry, reinforcement layout and detailing were accurately reproduced at model scales. The use of microconcrete and cold-rolled and annealed model reinforcement permitted an adequate simulation of basic material properties. Gravity load effect were simulated by placing lead weights at discrete points of the model specimens. All specimens were subjected to quasi-static cyclic loading which reproduced the displacement histories applied in prototype tests.

The purpose of the following discussion is to illustrate a few test results, assess the accuracy with which prototype behavior is reproduced in small-scale models, and present practical information that is relevant to the seismic design of frame - shear wall structures. In this discussion, all results obtained from the model tests are presented in the prototype domain.

BEAM-COLUMN ASSEMBLY TESTS

Typical results of beam-column assembly tests are presented in Figs.3 and 4. Figure 3 shows parts of the beam load versus beam deflection history for four specimens, one being a prototype specimen (Fig.3(a)) tested at the University of Texas at Austin, the other three being 1:12.5 scale model specimens. The results in Figs.3(a) to 3(c) were obtained from interior assemblies of the type shown in Fig.2, whereas those in Fig.3(d) were obtained from an exterior assembly in which the beam and slab on one side of the column were omitted.

The results of Figs.3(a) and 3(b) permit a direct comparison of model and prototype test results since the specimens are identical except for size and certain differences in material properties (the yield strength of the model slab reinforcement is 28 percent higher than that of the prototype slab reinforcement). The response of the two specimens is similar except for the higher strength of the model specimen in the positive (downward) loading direction. The main reason for this discrepancy is the difference in yield strength of the slab reinforcement. In both the prototype and model specimens it was found that the slab contributes much more to the beam strength than was anticipated. The nominal design strengths based on the ACI T-beam approach and on the measured yield strength of the reinforcement are indicated as P_n^{ACI} in the graphs. Also shown and denoted as P_n^s are the design strength values based on assuming that the full slab width is effective for beam bending. The experimental results show that the true strength values are closer to the latter than the former.

The importance of the slab in resisting bending can be seen by comparing the results of Fig.3(b) with those of Fig.3(c). The latter were obtained from a model specimen in which the floor slab was omitted. The much lower strength and its agreement with the predicted value are evident in the graph.

A notably different strength for positive loading is observed in the response of an exterior assembly (Fig.3(d)) in which the beam and slab on one side of the column were omitted. In this case the slab did not contribute nearly as much to the beam strength as in the interior assembly since the torsional flexibility of the transverse beams framing into the column did not permit mobilization of full slab action.

It must be concluded from these tests that the "beam" strength and stiffness of cast-in-place floor systems depends strongly on slab action and that the extent of slab action depends on the boundary conditions at the column which are different for interior and exterior joints. The shapes of the hysteresis loops may depend as well on these boundary conditions. In an interior assembly the reverse action of the two beams framing into a joint (tension in a rebar on one side of the joint and compression in the same rebar on the other side) may lead to bond deterioration in the joint and considerable pinching of the hysteresis loops (see Figs.3(a) to 3(c)). In an exterior assembly these pinching will be smaller provided the rebars are sufficiently anchored into the joint.

The fact that bond deterioration in the joint did occur in both the model and prototype interior assembly is illustrated in Fig.4. This figure shows readings from strain gages attached to the bottom reinforcement of a beam at the column face. Under loading that is expected to cause compression in the bottom bars (positive drift angle), both the model and prototype results indicate tension. The only logical explanation is that bond has deteriorated through the joint and the bar is subjected to tension because of the reversed moment applied to the opposite beam.

WALL AND WALL - FRAME TESTS

Test Results and Observations

The 1:12.5 scale models of the isolated wall and the wall - frame unit (shown in Fig.5) were loaded with a lateral load pattern similar to that applied to the full-scale structure in Tsukuba. The cyclic loading history was also modeled after the full-scale structure by reproducing the scaled roof displacement of the pseudo-dynamic tests performed on the Tsukuba structure.

In the isolated wall model, flexural cracks in the boundary elements of the shear wall were observed early in the loading history. At larger loads, these flexural cracks propagated as diagonal cracks across the shear wall, creating the characteristic cross pattern of diagonal tension cracks. As large displacements were imposed on the specimen, the diagonal crack pattern stabilized because the strength of the specimen was limited by flexural hinging at the base of the wall. This hinging was evident from large horizontal cracks in the boundary elements and a major crack at the base extending across the full wall. During the last cycle of the most severe test (PSD-4), crushing and bar buckling was observed in the boundary elements and several reinforcing bars fractured upon load reversal.

A somewhat different behavior of the wall was observed in the wall - frame specimen. The response of this specimen was affected strongly by three-dimensional wall-frame interaction. Because of this interaction, which is described in detail later, the wall of the wall - frame specimen resisted a relatively small portion of the base moment but a relatively large portion of the base shear generated by the lateral loading. Thus, shear cracking in the wall was more pronounced in this specimen than in the isolated wall specimen. Nevertheless, the strength of the wall in the wall - frame specimen was still governed by flexural hinging at the base. But since shear deformations contributed more to lateral displacements, the wall did not fail by rebar fracture during the displacement controlled test PSD-4 as did the isolated wall. Failure occurred at a very large displacement by crushing in a boundary element and in the adjacent corner of the shear wall.

Because of its effect on strength and failure mode, the three-dimensional wall-frame interaction became a focal point in this cooperative research effort. This interaction is caused by in-plane rotation as well as vertical displacement of the boundary elements of the shear wall. The effect of these deformation components is to mobilize not only the bending resistance of the in-plane beams framing into the wall, but also that of the floor slab and the transverse beams framing into the wall. It can be seen from free body diagrams that the three-dimensional wall-frame interaction will reduce the moment in the wall more than the shear in the wall and that it will increase considerably the axial force in the wall. The latter effect in turn will increase the bending strength of the wall.

All these effects appear to be beneficial since they serve to increase the overturning moment that can be resisted by the structure. However, it must be recognized that the wall-frame interaction will decrease severely the moment to shear (M/V) ratio in the wall and may change the failure mode from a ductile flexural hinging mode to a less ductile shear mode. For this reason this interaction should be considered in the design process.

The three-dimensional aspect of wall-frame interaction comes primarily from shear wall rocking that causes large extensions on the tension side of the wall but only small contractions on the compression side (see Fig.6). Because of the large extensions, plastic hinges may develop in the transverse beams framing into the tension boundary element and large shear forces are transferred from the transverse beams to the boundary element. At maximum lateral load these shears, which were measured in linkages restraining the transverse beams, resisted 11% of the base overturning moment applied to the test structure. The axial force generated in the walls by the beam shears increased the bending strength by about 20%. Thus, the effect of transverse beam action is to increase considerably the overturning moment resistance of the structure.

A large portion of the wall-frame interaction comes from frame action in the plane of the shear wall. Although this effect is usually recognized in design, it is apt to be underestimated because of the previously discussed large contribution of the slab to beam bending. This holds true particularly at large displacements at which many plastic hinges have formed in the beams. For the model test structure, this in-plane frame action at maximum load resisted 36% of the base overturning moment but only 9% of the base shear.

When all effects of wall-frame interaction are accounted for, it is found that the wall of the wall - frame specimen had to resist 91% of the base shear but only 53% of the base overturning moment. Furthermore, wall-frame interaction is predicted to increase the bending strength of the wall by 25% (20% due to transverse beam action and 5% due to in-plane frame action). Considering all these effects and noting that the wall of the wall - frame specimen failed in bending, it is predicted that the maximum lateral load (base shear) on the wall - frame unit is 1.25/0.53 = 2.36 times as large as that on the isolated shear wall. A strength increase close to this factor was observed in the tests as can be seen from Fig.7. This figure shows parts of the base shear versus roof displacement response of the isolated wall and the wall - frame unit.

Correlation with Full-Scale Test Results

The 1:12.5 scale model of the wall - frame unit represented that portion of the prototype structure that contributed most to structural strength and stiffness. The effects of interaction with the two exterior frame units that are not represented in the model were considered through adjustments in the applied loading pattern. Thus, the load-deformation response of the model structure should represent the response experienced by the center portion of the prototype structure.

The global damage patterns in the model specimen and the prototype structure tested in Tsukuba were indeed very similar. In both test structures considerable spalling of concrete was observed in the bottom of longitudinal beams at the wall boundary elements. Spalling was observed also in the boundary elements close to the base, but to a greater extent in the model specimen. Crack patterns in floor slabs, beams, boundary elements and shear wall were similar but with fewer cracks at wider spacings in the model specimen. Both the model and prototype specimens survived a rigorous test series that represented the effects of severe seismic ground motions without noticeable deterioration in strength.

In order to compare the overall load-deformation response, the model test results were modified analytically to account for the effects of the missing frame units. In this modification, advantage was taken of the experimental information generated in the beam-column assembly tests. An example of the modified model test results together with the corresponding prototype data is shown in Fig.8. This figure shows the base shear versus roof displacement response for the test PSD-3.

CONCLUSIONS

The model experiments performed as part of the U.S.-Japan cooperative research program have demonstrated that the overall response of reinforced concrete components and structures can be reproduced well in tests with small-scale models.

The engineering lessons learned from the small-scale model tests were the same as those learned from the full-scale tests performed in Tsukuba. The major lessons learned are the following:

- The test structure behaved well under cyclic loading that represents the effects of severe seismic ground motions.
- The floor slab contributes strongly to the effective strength and stiffness of the beams in cast-in-place reinforced concrete frame - shear wall structures.
- Three-dimensional wall-frame interaction has a strong effect on the overturning moment resistance of a frame - shear wall structure. Again, the floor slab contributes much to this effect. The use of two-dimensional analysis and of ACI T-beam strength would severely underestimate the true wall-frame interaction.

ACKNOWLEDGEMENTS

This research was funded by the National Science Foundation through the Grant CEE 80-21119. This support and the skillful assistance of the graduate students R T Sewell and E L Tolles are gratefully acknowledged.

Figure 1 Plan view and elevation of prototype structure

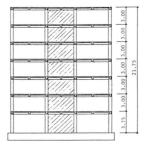

Figure 2 Beam-column assembly test specimen

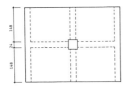
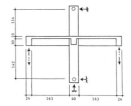

Figure 3 Beam response in beam-column assembly tests

(a) Interior assembly with slab
(prototype specimen)

(b) Interior assembly with slab
(1:12.5 model specimen)

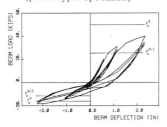
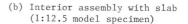
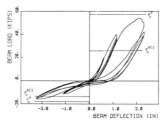

(c) Interior assembly without slab
(1:12.5 model specimen)

(d) Exterior assembly with slab
(1:12.5 model specimen)

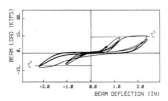
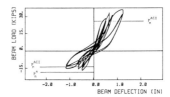

Figure 4 Strains in beam bottom reinforcement of interior assembly
(a) Prototype specimen   (b) Model specimen

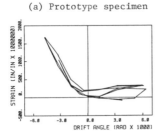
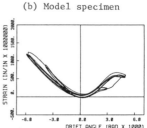

Figure 5 Shear wall - frame model   Figure 6 Vertical displacement of
                                                wall boundary element

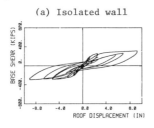
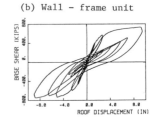

Figure 7 Base shear vs roof displacement response of model specimens
(a) Isolated wall   (b) Wall - frame unit

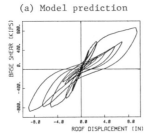
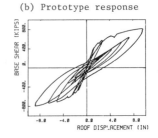

Figure 8 Predicted and observed response of complete structure
(a) Model prediction   (b) Prototype response

DISCUSSION

Dr M Dostal   National Nuclear Corporation   UK

In your conclusion (as presented but not included in the written paper) you stated that the model structures proved to be much more stronger than anticipated. Could you please enlarge on this, in numerical terms if possible. Is this conclusion relevant to realistic structures or is it explained by peculiarities of the test models?

Author's Reply

This discussion addresses the conclusion drawn in the paper that the strength of the model test structure was much higher (about 50% higher) than was predicted by analysis. The high strength of the model test structure is not at all attributed to model similitude problems; in fact, the strength of the model structure was very close to that of the prototype structure tested in Japan. The issue is that conventional analysis techniques are simplistic and ignore several phenomena that contribute to strength.

To place this conclusion in perspective, it must be emphasised that the strength of the test structure was governed by bending resistance and not shear resistance of the shear wall. Sufficient shear reinforcement was provided in the wall so that a flexural hinge formed at the base of the wall before a shear failure mechanism developed. Since bending strength controlled the behaviour, the quality of an analytical prediction depends on the accuracy with which the overturning moment resistance at the base of the structure can be modelled.

The differences between the measured strength and the strength predicted from conventional analysis are attributed to the following three sources:

1. The bending strength of the "beams" is likely to be severely underestimated. All tests of the co-operative research program have demonstrated that the floor slab contribution to "beam" strength increases considerably in the inelastic range. At large beam rotations a major portion of the floor slab becomes effective in resisting bending.

2. The vertical elongation in the tensile boundary element of the shear wall causes high moments and shear forces in the beams framing transversely into the shear wall. This outrigger action of the transverse elements framing into the wall contributes to the bending resistance of the structure.

3. The shear forces in the transverse beams as well as in the longitudinal beams are transferred as axial forces into the shear wall. For walls of normal proportions, these additional axial forces will increase the bending resistance of the wall.

The importance of these three items is illustrated in the following example, using the prototype test structure and simple limit analysis for illustration. Assuming a triangular lateral load pattern and a complete mechanism for the two frame units and the frame - shear wall unit of the structure, a maximum overturning moment resistance of approximately 33,000 kip-feet is predicted from conventional analysis. This corresponds to a base shear resistance of 640 kips. Conventional analysis implies here that the beam strength is predicted from the ACI T-beam approach and that the three aforementioned effects are not considered. If the analysis is repeated with the three effects included, a 53% increase in the lateral limit is obtained. From this increase, 36% are attributed to the first effect, 8% to the second effect, and 9% to the third effect.

This example shows that the increase in beam strength due to slab contribution appears to be the predominant effect. At this time it is hardly possible to predict accurately the beam strength including slab contribution. In the analysis example the measured strengths obtained from component tests were used for this purpose. It is worthwhile to note that the base shear resistance so calculated (1.53 x 640 = 980 kips) compares well with the maximum resistance measured in the prototype test. The base shear versus roof deflection diagram for the prototype test in which the maximum resistance was attained is shown in Fig 1.

The concern about the maximum resistance of a frame - shear wall structure may appear to be only of academic value. Our contention is that this is not true and that there is a hidden but important design implication. In earthquake resistant design it has long been recognised that flexural hinging of a shear wall is a much more desirable limit state than shear failure which is often of brittle nature. Thus, shear walls should be designed so that they hinge in bending rather than fail in shear. In order to fulfil this objective it will be necessary to predict accurately the maximum lateral load resistance of a structure so that sufficient shear resistance can be provided in a wall to avoid shear failure.

Figure 1 Prototype response -- Test PSD-4

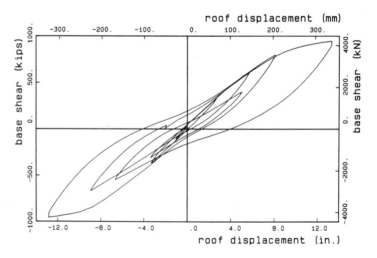

28 Reinforced concrete beam-column subassemblages under reversed loading

E. C. CARVALHO and S. POMPEU SANTOS
Laboratório Nacional de Engenharia Civil, Portugal

Summary: The paper describes the recent experience of the National Laboratory for Civil Engineering (LNEC) in the experimental study of reinforced beam-column subassemblages under reversed loading. Three series of tests with different objectives are described. The tests set up and the techniques used are referred. Some conclusions and comments are also presented.

1 - INTRODUCTION

In seismic areas, structures are designed so that in the event of a large intensity earthquake they should respond well beyond its initial elastic behaviour. In this way it is expected that large amounts of the energy input by the earthquake into the structure may be dissipated by hysteresis. Such dissipation is dependent on the type of nonlinear behaviour under large amplitude reversed cycles developed in some critical regions of the structure. Wide, non-degrading restitution loops are needed for such purpose and thus, sources of brittleness and of stiffness and strength degradation must be avoided. Although there are already analytical models (1) available for the treatment of some specific situations, experimental work is still needed in many circumstances for the assessment of the nonlinear hysteretic behaviour of reinforced concrete structures or substructures. The purposes of such tests may be quite different even when conducted on similar types of elements with similar experimental techniques.

LNEC had recent experience in this field, in connection with three sets of tests performed in reinforced concrete beam-column subassemblages subjected to static reversed loads (imposed displacements) simulating earthquake actions. The different objectives of each set of tests were:

a) Research on the ultimate strength and deformability of beam-column joints;
b) Check of the available ductility of a framed structure with tests conducted on 1:2.5 subassemblages directly reproducing the real design;
c) Research on the seismic behaviour of reinforced concrete precast framed structures with tests conducted on 1:1 beam-column subassemblages.

Due to the nature of this paper only summarised references about tests are made.

2 - GENERAL DESCRIPTION OF THE TESTS

All the three sets of tests were conducted in external beam-column subassemblages which represent a part of a framed structure for which it is acceptable to consider zero bending moments values in sections at mid-span of its members. The models, T shaped, were placed vertically in a testing rig available at LNEC. Columns ends were simply supported by means of adequate steel bearings which restrained horizontal displacements but permited rotations. A constant axial load was applied in the column by a manual hydraulic jack. Alternated or repeated displacements were applied by hydraulic jacks at the end of the beam following "a priori" defined amplitude sequences. A general view of the test set up for one of the experimental series is presented in Fig. 1.

Figure 1 - General view of tests set up

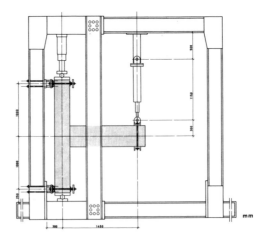

3 - ULTIMATE STRENGTH AND DEFORMABILITY OF BEAM-COLUMN JOINTS

In framed concrete structures beam-column joints behaviour is of great importance. In order to assess the structural behaviour of the joints under seismic actions, an experimental program was carried out utilising concrete models of external beam-column subassemblages.

Three pairs of models (CX2NA, CX2A and CX4NA, CX4A and CX8NA, CX8A) of the same geometry were tested. Columns were 0.85 m long and had 0.20 x 0.20 m cross sections. Beams were 1.25 m long and had 0.20 x 0.30 m (width x height) cross sections. Main reinforcement of the elements were identical, but transversal reinforcement at joint region was different. In Fig.2 a general view of the models is presented.

Figure 2 - General view of models of the 1st serie tests

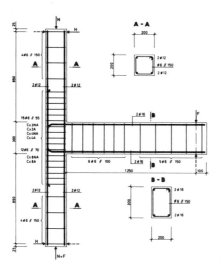

Models "NA" were tested with repeated loads while models "A" were tested with alternated loads, for comparison. The alternated displacement sequences depends on the joint behaviour but increased from cycle to cyle. The column axial compression loads represents about 17%, 35% and 70% of concrete resistance of models "2", "4" and "8", respectivelly. Instrumentation of the models mainly consisted of inductive and mechanical displacement transducers placed in such way to obtain, among others, displacements at the beam end and rotations between cross sections in beam and column critical regions. In

Fig. 3 two diagrams relating the ratio of the bending moment in beam at column face to the ultimate (theoretical) value, $M_v/M_{r,v}$ (normalised bending moment) is presented in function of the displacement at beam end. More detailed information about these tests may be found in (2).

Figure 3 - Normalised bending moment - Displacement diagrams from two models of the 1st serie tests

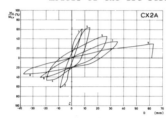 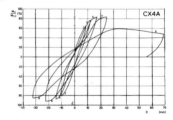

The basic conclusions obtained by the author were:

i) Shear for first crack at joints can be well predictable by Taylor formulation (3).
ii) Behaviour of joints confirms the theoretical model that considers the concrete strut and the struss mechanism for joint shear resistance.
iii) Alternated loads produces a deterioration of the strut mechanism of the joints, particularly for low axial loads in columns.

4 - AVAILABLE DUCTILITY IN A FRAMED STRUCTURE

In order to assess the available ductility in a framed structure designed for a large urban development, an experimental program was carried out. Typically the buildings were five to nine stories high and its structures were reinforced concrete moment-resisting space frames with two principal orthogonal directions. Girders spans were either 8.25 x 8.25 m or 11.0x5.5m supporting quadrangular or rectangular waffle slabs. The tests were requested to LNEC by the structural designer who was particularly concerned with the seismic behaviour of the beam-column joints. As many different situations can occur in the structure, the choice of typical situations were required for testing purposes. Although internal joints might represent more severe situations, the short time available for the tests led to the choice of external joints whose tests could be conducted in the readily available experimental facilities.

Two pairs of models (1A, 1B and 2A, 2B) were constructed reproducing, directly from the design drawings of a typical frame, lower and upper stories beam-column subassemblages. The geometrical scale chosen was 1:2.5 and, as far as materials were concerned, a direct reproduction of their mechanical properties was envisaged. Columns were 0.51 m long and had 0.26 x 0.26 m cross sections. Beams were 1.52 m long representing (to scale) half span of the frame. Cross sections were for models 1A and B, representing the lower stories, 0.22 x 0.26 m and for models 2A and 2B, representing the upper stories, 0.22 x 0.28 m. The shear ratio was thus of the order of 6, which allowed to predict, for the beam, a predominance of the flexural behaviour. In Fig. 4 a general view of models 1A and 1B is presented.

Figure 4 - General view of two models of the 2nd serie tests

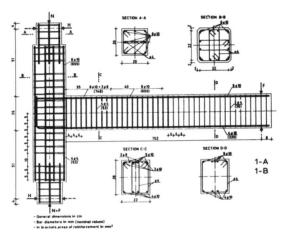

Except for model 1B, the amplitudes of the alternated displacements sequences increased from cycle to cycle although in an asymmetrical way due to the differences between top and bottom reinforcement in the beams. For model 1B an irregular pattern was adopted based on a random application of the cycles amplitudes. In model 2B a constant force scaling the effect of the permanent gravitational loads in terms of bending and shear in beam,was also applied. Fig. 5 shows the applied force and the bending moments diagrams pattern of the models. Instrumentation of the models mainly consisted of several inductive transducers placed in such a way as to obtain, among others, the following diagrams: Force-Displacement at the beam end, Force-Relative rotation between adjacent cross sections (20 cm apart)

Figure 5 - Simulation of the gravitational loads in beams of the 2nd serie tests

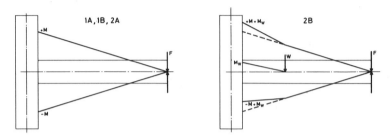

in the beam critical region, and Force-Distortion of the beam-column joint. Force-Displacement diagrams obtained are presented in Fig. 6. More detailed information may found in (4) and (5).

Figure 6 - Force-Displacement diagrams from models of the 2nd serie tests

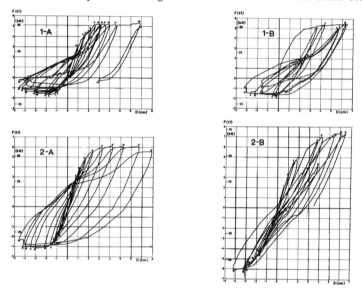

In all the four models plastic hinges developed in the critical region of the beam. Ductility available and energy dissipation capacity were good and enough to develop the nonlinear behaviour considered in design of the real structure. As the models reproduced part of the "as designed" structure, beams had unsymmetrical reinforcement and curtailments at its top bars. These features, combined with the application of alternated displacements (forces) at the beam created, for models 1A, 1B and 2A, a clearly different behaviour for forces downwards and upwards which obviously is not representative of an earthquake response, that should be essentially symmetric. For model 2B ("equal" to model 2A) for which the simulation of the bending and shear in the beam due to the vertical loads was made, the diagram F-D obtained became in fact pratically symmetric in spite of the unsymmetric reinforcement (and the curtailments).

5 - SEISMIC BEHAVIOUR OF REINFORCED PRECAST CONCRETE FRAMED STRUCTURES

The seismic behaviour of precast structures depends mainly on the ductility of the connections located at joint zones. In order to assess the available ductility of a type of r.c. moment-resisting precast structure commonly used in Portugal, an experimental program was carried out on models reproducing the job conditions of the real precast structures. These structures use precast elements in columns and in beams completed with concrete in field. Connections are obtained with reinforcement bars anchored at the joint core or at the column core.

Two pairs of models (PVP1, PVP2 and PVP3, PVP4) of identical geometry were constructed, reproducing two typical situations of relative resistance of beams and columns. Columns were 0.80 m long and had 0.30 x 0.30 m cross sections. Beams were 1.30 m long and had 0.15 x 0.40 m cross sections. Thus, the shear ratios of columns and beams were of the order of 7 and beam-column length ratio was about 3 to 2, as current in real structures. In fig. 7 a general view of the models is presented.

Models PVP1 and PVP3 were tested with repeated loads while models PVP2 and PVP4 were tested with alternated loads, for comparison. The alternated displacements sequences were symmetrical and increased from cycle to cycle. The column axial compression loads represents about 11% and 5.5% of concrete resistance of the pairs of models PVP1, PVP2 and PVP3, PVP4, respectivelly. Instrumentation of the models mainly consisted of several inductive transducers placed in such a way as to obtain, among others, displacements at the beam end, rotations between cross sections in beam and columns connections, and rotations and distortions of the joint, which were directly recorded versus the force applied at beam end. The Force-Displacement diagrams at beam end obtained from models PVP2 and PVP4 are presented in Fig. 8. More detailed information about these tests may be found in (6).

Figure 7 - General view of models of the 3rd serie tests

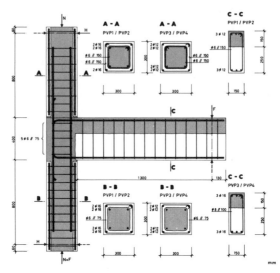

Figure 8 - Force-Displacement diagrams from two models of the 3rd serie tests

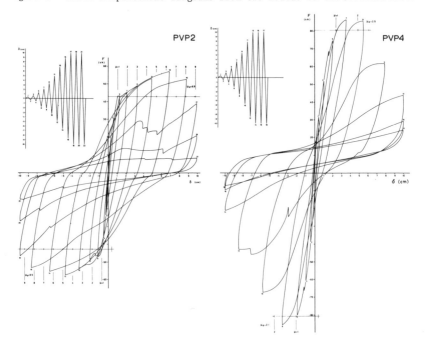

The basic conclusions obtained were:

i) General behaviour of beam and column connections and joints is similar to the monolithic reinforced concrete elements. Joint resistance will be obtained neglecting the cover of joint stirrups.

ii) Ductility of beam-column region depends on the relative resistance of beam, column and joint. When plastic behaviour is concentrated in beams reinforcement, ductility may be large identical to the monolithic beams. When plastic behaviour is concentrated in columns reinforcement, ductility decreases because cracks concentrate at columns base section. When plastic behaviour is concentrated in joint stirrups, ductility becomes very low because a shear failure in the joint is obtained.

6 - FINAL REMARKS

Results obtained in the above described tests show the great interest of experimental techniques to solve problems involved. Nevertheless, on the application of these techniques some features must be taken in account. So, to study a specific problem it's necessary to choose carefully the parameters analysed, to reduce, as many as possible, the number of tests, because they are normally very expensive. In tests conducted in models simulating some kind of reality, attention must be paid to the need to simulate all the relevant conditions involved in such reality.

REFERENCES

(1) CEB-Comité Euro-International du Béton, Response of R.C. Critical Regions Under Large Amplitude Reversed Actions, Bulletin d'Information Nº. 161, Paris/Lausanne, (August 1983).

(2) NASCIMENTO, J S, Comportamento de Conexões Viga-Coluna de Betão Armado Submetidas a Esforços Repetidos", Thesis (in Portuguese), LNEC, (1977).

(3) TAYLOR, H J, The Behaviour of in Situ Concrete Beam-Column Joints, Technical Report, Cement and Concrete Association (CCA), May, (1974).

(4) CARVALHO, E C and RAVARA, A, Ensaios Sísmicos em Modelos das Estruturas das Áreas Centrais de Morelos, La Hoyada e Carabobo, em Caracas, Final Report (in Portuguese), LNEC, Lisbon, (1979).

(5) CARVALHO, E C, Beam-Column Subassemblages Tested Under Alternated Loads for Seismic Design, AICAP-CEB Symposium, Structural Concrete Under Seismic Actions, Comité Euro-International du Béton, Bulletin d'Information Nº. 132, Paris (April 1979).

(6) SANTOS, S POMPEU, Comportamento de Ligações de Estruturas Prefabricadas de Betão, Thesis (in Portuguese), LNEC, Lisbon, (1983).

29 Use of physical silo models

J. MUNCH-ANDERSEN and J. NIELSEN
Department of Structural Engineering, Technical University of Denmark

SUMMARY

Physical silo models have been used at the Department of Structural Engineering to investigate flow patterns and pressure distributions. Both centrifuge models and models in the natural field of gravity have been used. The model laws show that many assumptions have to be fulfilled if reliable and accurate results are to be obtained. Two applications are referred to: a pressure measurement on a grain silo and a model of a silo for crushed oyster shells.

The grain silo investigation consists of full-scale wall pressure measurements checked by measurements in both types of model. Despite careful fabrication and an advanced pressure cell technique, serious scale errors have been ascertained in some cases. It is concluded that many silo model tests are carried out without sufficient proof of reliability.

The model tests with crushed oyster shells were made in order to arrive at a silo shape that would firstly secure flow of the medium and secondly cause a specific flow pattern (mass flow). There is no existing theory to cover this very special medium, which consists of disc-shaped particles. The centrifuge testing led to the development of a design, and a full-scale silo was built on this basis. However, although flow did occur, the flow pattern was not as intended - probably due to poor finishing work (surface too rough) on the full-scale bottom cone.

It is concluded that important results can be obtained from silo models but that there is a serious risk of misinterpretation.

INTRODUCTION

Designers of silo structures are faced with some complex problems:

The overall design of a silo, especially the shape and the roughness of the bottom, determines the flow conditions, which can be very important to the working conditions of the silo. High costs as a consequence of flow stop etc. have been reported. For the most simple bottom shapes theoretical solutions are available for homogeneous, isotropic powders. Tests are still necessary for more complicated silo shapes and for silos for special media. An example with a highly anisotropic medium, crushed oyster shells, is discussed.

The structural analysis is based on load assumptions taken from the codes of practice. These codes are mainly based on practical experience and results from physical models. Consequently, models are often used when new knowledge is needed. Determination of scale errors in such tests is therefore important to the reliability of the codes. An example with grain silos is included in this paper.

Finally, the silo structure itself has to be analysed. Such analysis can be very complicated when dealing with shell structures built together, but since the development of computer techniques, physical models play a smaller role in this part of silo design.

MODEL TECHNIQUE

Two types of silo model are discussed in this paper, a large model in the natural field of gravity and a small centrifuge model.

For models in the natural field of gravity the pressures are scaled almost as the length. The model law requires the same angle of internal friction of the medium, coefficient of friction along the wall, and strain, in the model as in the prototype, and the same scale as the length scale for the modulus of elasticity of the medium and the wall material, and for the cohesion, if any.

According to the model law, the particle size must be in the same scale as the model, which requires a substitution of the medium. Centrifuge tests have revealed that such substitution is almost impossible. Therefore, all model tests are carried out with the prototype medium.

When the centrifuge technique is used with the prototype medium, very few assumptions are necessary because the true stress level and strain field can be obtained. On the other hand, however, the coriolis-forces and the grain size give rise to problems (1), (5).

GRAIN SILOS

Alarming cracks are often observed in high silos of reinforced concrete. Usually, the horizontal pressure during discharge is regarded as the problem. A research project covering pressure measurements in a full-scale silo and in a 1:10 model in the natural field of gravity and an investigation of the material behaviour (barley and wheat) has been carried out by the "Nordic group for silo research" (2), (3).

The material investigation revealed that the stiffness of grain is almost proportional to the consolidation pressure and that the apparent cohesion tends towards zero at low pressure, meaning that the requirements of the model law are reasonably satisfied.

The full-scale silo has a height of 45 m and a diameter of 7 m. The 1:10 model is made of epoxy, the stiffness of which is close to 1/10 of that of reinforced concrete, as required by the model law. Both silos are equipped with pressure cells built into the wall and covered so that the stiffness, roughness, and geometry, of the wall remain unchanged. The importance of this is discussed in (1).

During discharge, different types of flow can occur, as shown in Fig. 1.

Figure 1. Flow patterns

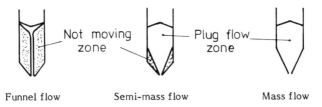

Funnel flow   Semi-mass flow   Mass flow

During semi-mass flow a boundary layer adhering to the wall is observed in concrete silos, revealing that the effective coefficient of friction is determined by the grain-to-grain friction.

The thickness of the boundary layer is several times the grain size. Preliminary tests in a smooth model silo without a boundary layer revealed systematic deviations from the expected symmetry in the pressures, even though great care was taken to minimize the imperfections, which are known to influence the pressure distributions (1).

The expected symmetry was obtained when a boundary layer occurred after the inner surface had been roughened by gluing a layer of sand to the wall.

The boundary layer is relatively thinner in the full-scale silo, and the imperfections cause systematic deviations from a symmetrical pressure distribution.

The type of inlet greatly influences the densities (i.e. the strength) and thus the pressure at rest and the type of flow, as shown in Fig. 2.

Figure 2 Observations in model and full-scale barley silos.

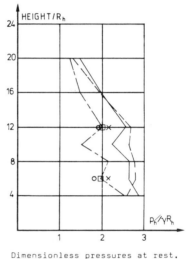

INLET TYPE		SCALE	DEN-SITY	FLOW	SIGN.
A		1:1	8.0	F→SM	———
		1:10	7.0	SM	×
B		1:1	7.9	SM	— —
		1:10	7.5	SM→F	□
C		1:1	8.3	F	—·—
		1:10	7.7	F	○

Observed densities (in kN/m^3) and flow patterns (SM=semi-mass flow, F=funnel flow).

Dimensionless pressures at rest.

According to the theory, the asymptotic, horizontal pressure should be $p_h = \gamma R_h/\mu$ where γ is the density, R_h the hydraulic radius of the silo, and μ the coefficient of friction between the wall and the grain.

The differences in the dimensionless pressures at rest are caused by the different strength and thus different μ. In the model, the different densities fully explain the differences in pressure, but in full scale, the inlet types A and B cause unexpectedly large pressures. The reason for this is that during the beam-filling in full scale, some sliding layers form on the top cone, which reduces the strength in narrow zones. This strength seems to influence the global strength (4). In the model, such layers are avoided, presumably due to the bigger ratio of grain to silo diameter.

During funnel flow the discharge pressures are very close to the rest pressures, but when semi-mass flow occurs, the pressure changes considerably. This makes the flow pattern important to the safety of the silo, so it is important to emphasize that flow and density are related in the same way in both scales, as shown in the table in Fig. 2.

Figure 3 Examples of pressure versus time curves in model and full scale.

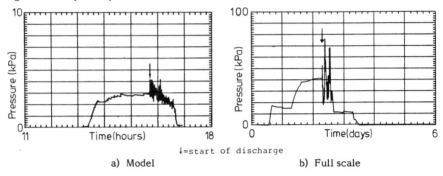

a) Model ↓=start of discharge b) Full scale

After start of discharge the pressure in the model fluctuates around a level that is greater than the rest pressure (Fig. 3a), while it varies slowly in full scale (Fig. 3b). In the full-scale tests, filling and discharge were interrupted during the night, as indicated in Fig. 3b. The average pressure remains unchanged in full scale.

The fluctuations in the model are local (but not dynamic). In full scale, there is a global redistribution of the pressure, causing serious bending moments in the silo wall - a loading case which is not considered in theories and codes of practice (except in the latest edition of the German Code, DIN 1055).

The necessary dilation in the boundary layer gives rise to an overpressure of the order of magnitude of 1 kPa, independent of the scale, thus explaining the increased pressure in the model during discharge.

In full scale, the uneven pressure distribution can be explained by a combination of the imperfections and the weak zones caused by the bedding.

With rape seed, comparative tests have been carried out in an epoxy model and a small centrifuge model. The small rape-seed grain (dia. 1.3 mm) allows sliding layers to form on the surface in the larger model, but not in the centrifuge model. The tests confirm many of the conclusions mentioned above.

SILO FOR OYSTER SHELLS

For a Danish company, Hotaco A/S, which sells crushed oyster shells, a 6 m diameter and 21 m high silo, as shown in Fig. 4, was proposed. The silo had to be able to handle different fractions of crushed oyster shells (used for chicken feed).

Crushed oyster shell particles are disc-shaped, having a thickness which is typically one order of magnitude less than their dimensions in other directions. Usually, such particles stack in such a way as to form a highly anisotropic medium (4), for which no silo theory can be used. Standard triaxial tests carried out at the Danish Geotechnical Institute confirmed that the material could not be described by the normal soil parameters.

For that reason and because this silo is much larger than oyster shell silos built earlier (1978), it was decided to investigate the project by model testing. The main questions were:

Figure 4 Initial design

Figure 5 Centrifuge model. Cylindrical part different cones, and inserts

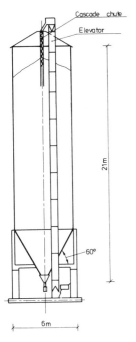
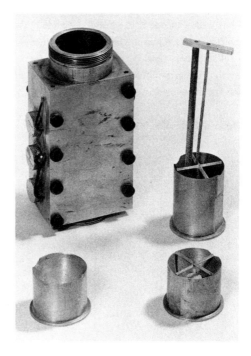

- Will the oyster shells run out through the outlet by gravitation alone or is flow promotion equipment necessary?
- Will the flow pattern be mass flow, which is attractive from the point of view of separation and breakdown of the particles?
- Will the inside elevator significantly influence the pressure distribution?

As different bottom shapes and inserts had to be tested and because it was attractive to perform the tests as close to the right stress level as possible in order to avoid assumptions concerning material behaviour, it was decided to make centrifuge tests in spite of lack of experience with the scale errors for centrifuge silo models (5). Unfortunately, the maximum speed of the centrifuge corresponds to a 75 times increase in the field of gravity, whereas 150 times is called for according to the model law. As the best approximation, the measured pressures were multiplied by two, while the flow pattern was assumed to be correct.

The finest fraction of oyster shells (pass an 0.5 mm sieve) was used for the test because it was considered to be the critical one.

Tests with a model of the initial design (60° steel cone) showed that flow did occur, but that the flow pattern was the less attractive funnel flow (6). Tests with a 70° cone did not show a reliable mass flow. Different kinds of inserts (see Fig. 2) improved the flow conditions, and because the alternative - a steeper or smoother cone - was expensive, a 70° cone with insert was recommended.

Figure 6 Main dimensions of the model and pressure versus time curves for three different cell positions relative to the elevator position (6)

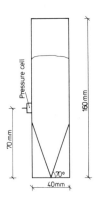
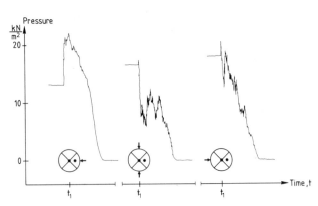

Finally, a test series was run to check the elevator's influence on the pressure distribution. A photo of the model of the elevator and the cascade chute is shown in Fig. 5. Pressure versus time curves can be seen in Fig. 6. Here, it will be seen that the pressure distribution at the level in question is very far from rotational symmetrical. During rest ($t < t_1$), the pressure is small at the part of the wall nearest the elevator. Shortly after start of discharge, the pressure is big in directions corresponding to a diameter through the elevator and small in a direction perpendicular to this. Such pressure distributions will lead to considerable bending moments in the concrete wall and proably also to a horizontal load on the elevator itself. As a result it was decided to place the elevator outside the silo cell.

The silo, which was built after these tests, has its cylindrical walls and the upper two thirds of the cone in concrete, while the lower part of the cone is made of steel and has an insert. Unfortunately, the surface of the upper part of the cone was made with a considerably greater roughness than prescribed in accordance with the model tests, and mass flow did not occur. Friction testing of a piece similar to the wall material confirmed that the difference in flow pattern could be explained by the roughness.

CONCLUSIONS

Physical models play an important role in silo design, but reliable results can only be expected provided great care is taken to fulfil the model requirements.

Grain silo models give good results in respect of flow behaviour and, in some cases, in respect of pressure distributions. Important deviations are explained by unsatisfied model requirements that are not commonly known and that therefore contribute seriously to the risk of misinterpretation of model tests.

A centrifuge model of a silo for crushed oyster shells has been used to reveal drawbacks in a proposed design and to develop a design with better flow and pressure conditions.

The final conclusion is therefore that important results can be obtained from models, but that the tests have to be performed with care and interpreted with caution.

REFERENCES

(1) ASKEGAARD, V, and NIELSEN, J, 'Measurements on silos', BSSB/ICE Joint Conference, Measurement in Civil Engineering, Newcastle, England, (5-8 September 1977).

(2) HARTLÉN, J et al., 'The wall pressure in large grain silos', Swedish Council for Building Research, Document D2, (1984).

(3) MUNCH-ANDERSEN, J, 'Scale errors in model silo tests', International Conference on Design of Silos for Strength and Flow, Stratford-upon-Avon, England, (7-9 November 1983).

(4) NIELSEN, J, 'Load distribution in silos influenced by anisotopic grain behaviour', International Conference on Bulk Materials Storage, Handling and Transportation, Newcastle, Australia, (22-24 August 1983).

(5) NIELSEN, J, 'Centrifuge testing as a tool in silo research', Symposium: The Application of Centrifuge Modelling to Geotechnical Design, Manchester, England, (16-18 April 1984).

(6) KOEFOED, K M, and KRISTIANSEN, N Ø, 'Forsøg med silomodeller (Tests with silo models)' (in Danish). M.Sc. thesis, Department of Structural Engineering, Technical University of Denmark, (1978).

30 Static tests on a 1:10 scale model of a reinforced concrete chimney

ALDO CASTOLDI
ISMES, Bergamo, Italy
GABRIELLA GIUSEPPETTI
ENEL-CRIS, Milan, Italy
LUIGI RUGGERI
ISMES, Bergamo, Italy
FIORE ULIANA
ENEL-CTN, Milan, Italy

1. SUMMARY

An 1 : 10 scale model of a reinforced concrete chimney (height 220 m, lower diameter 22 m, upper diameter 17 m) for a thermoelectric power plant has been tested at ISMES on behalf of ENEL (Italian State Electricity Board) (Fig.1.). This type of structure called for a more detailed stress analysis as against similar structures, since the holes provided near the chimney base, which make it possible for flues to pass, had been planned side by side rather than superposed, thus reducing the resisting section. The tests aimed at determining the stress distribution for dead load and design wind load, and the crack patterns and safety coefficient for wind load values increased up to failure. Special attention has been devoted to the determination of stress concentration around the holes set at the lower part of the structure. For this reason, only the lower 35 m (at prototype) have been reproduced, thus reducing model size and costs. The results obtained by using the physical model were compared with those achieved by a finite element mathematical model.

2. PHYSICAL MODEL

2.1 Design criteria

The geometric scale of the model was determined on the basis of previous experience suggesting the value λ = 10; this bearing in mind the model size and wall thickness, as well as the availability of equipment.

Reinforced microconcrete was used to build the model. Special attention was devoted to the choice of quantity and to the placing of reinforcement (Fig.2.). It was deemed advisable to give up the strict reproduction of wall reinforcement on geometric scale, resorting to the solution that devised the grouping of about 4 bars ⌀ 2.6 mm, into a single bar ⌀ 5 mm of equivalent section and placed at the centroid of the group itself. Mechanical properties of steel used for reinforcing the model wall and those of the prototype were the following:

Figure 1   View of reinforced concrete chimney model

Figure 2   Reinforcement placed in the model

	on the model	on the prototype
Young's modulus E_s	210,000 MPa	210,000 MPa
Yield strength f_y	600 MPa	440 MPa

Since the yield strength of the model reinforcement, consisting of small diameter bars, exceeded that of the prototype, it has been considered more correct to use as force scale that resulting from strength ratio rather than that ensuing from elastic moduli. This choice appeared reasonable because the chief aim of the model was the correct appraisal of the failure behaviour. The relevant force scale was: $S_F = \dfrac{440}{600} \times 10^2 = 73$.

As regards the reinforcement of the portion between holes, the thick stirruping with bars of very small diameter, as devised by the rigorous fulfilment of the geometric scale, was carried out by using 2 mm-diam stirrups, and by adopting an analogous grouping criterion. Moreover, since the yield strength of these stirrups was f_y = 850 MPa, to con-

Figure 3   Reinforcement of a portion between the holes

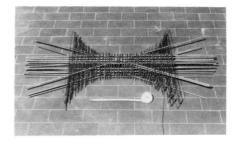

form the effect of this reinforcement to the force scale, a lower cross-sectional area had to be placed (Fig.3.).

The composition of cement mix and the grading curve of aggregates are shown in Fig.4. Max diameter of aggregates, Dmax = 8 mm, was fixed by taking into due account some factors, such as thickness of model wall, and spacing between reinforcement. Mechanical properties of micro-

Figure 4   Composition of micro-concrete

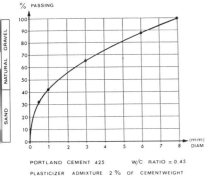

PORTLAND CEMENT 425   W/C RATIO = 0.45
PLASTICIZER ADMIXTURE 2% OF CEMENTWEIGHT

concrete at the beginning of tests (90 days curing) and those of the prototype were the following:

	on the model	on the prototype
Young's modulus E_c	30,000 MPa	30,000 MPa
Compressive strength f_{c_k}	50 MPa	35 MPa

It may be pointed out that the scale of force was quite observed also for the microconcrete ($S_F = \frac{35}{50} \times 10^2 = 70$).

2.2 Construction criteria

Wooden forms were first constructed for pouring the model. After the inner form was positioned the reinforcement was placed with the aid of appropriate spacers. The external form was subsequently placed. This consisted of superposed rings, making it possible to pour and compact in successive stages the microconcrete. The immediate succession of pours ensured the homogeneity of the model body. After the pour was completed, suitable devices were used, to make possible a good curing of the cement mix of the model for a period of some 90 days.

2.3 Carrying out of tests and instrumentation

Load conditions to be applied to the model had to reproduce the effect of dead load and wind action, the latter both in orthogonal and parallel direction to the hole axis. Dead load, bending moment and shear action were applied to the top of the model (El. 3.55) their values being such as to reproduce the design values at El. 1.555. A rigid r.c. plate was placed on top, forming one piece with the model. A set of hydraulic jacks was set on it; they loaded the model finding reaction, by means of appropriate equipment, on the chimney basement (Fig.5.). Dead weight was obtained through a compression on the model. The bending moment was reproduced by loading two diametrally opposite jacks, acting on the plate by opposite-sign forces. Shear action was reproduced as a side thrust reacting on a structure fixed to the basement. Reaction structure together with jacks reproducing the moment could be rotated around the chimney axis, making possible the rotation of the bending plane. Test to failure foresaw the increment of the wind effect only, in a direction parallel to conduit axis, the vertical load being kept at its constant value. Instruments used for stress measurement consisted of inductive extensometric rosettes, eletric resistance strain gauges, mechanical strain gauges. Their position on the model is shown in Fig.6.

Figure 5   Scheme of loading devices

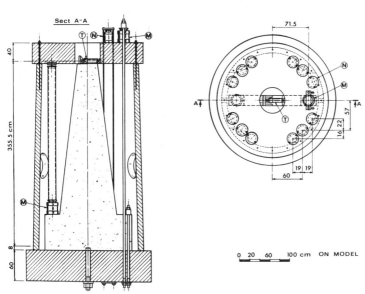

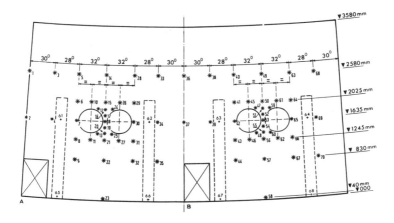

Figure 6   Instruments position on the model

A more accurate and thick instrumentation was placed at the portion between the holes, since it was the weakest part of the structure, and therefore it called for a wider investigation. The reading of instrument signals and their storage was performed by means of automatic data logger.

2.4 Test results

The stress state of the model, in the linear-elastic field, was evaluated from strain values measured by strain gauges and from elasticity parameters measured on microconcrete. Results are shown in Figs.7.a.b.c. in which amounts and directions of principal stresses in the various load conditions are plotted on the developed chimney external walls.

Test to failure was performed by increasing wind action in a direction parallel to conduit axis. A load of an intensity equal to 4.4 times that of the design was attained. At this load, a single crack occurred at the basement-wall connection, where max tensile stresses were produced, while around the holes no damage was observed.

Figure 7.a   Dead load condition: principal stresses

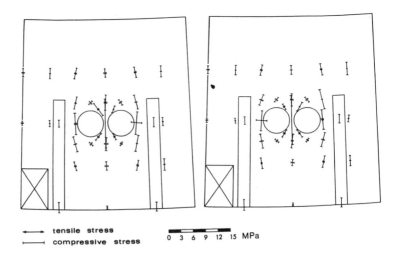

Figure 7.b   Wind action in a direction parallel to hole axis

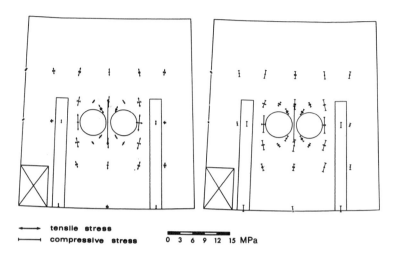

Figure 7.c   Wind action in a direction perpendicular to hole axis

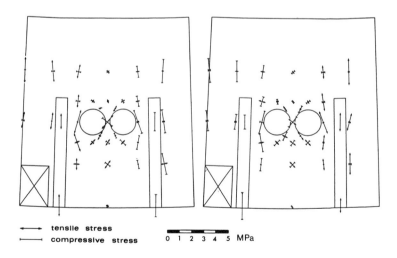

3. MATHEMATICAL MODEL

The study of the stress-strain state in the linear-elastic behaviour relevant to the lower part of the chimney including base openings and flues inlet was developed, in parallel with physical model, by a three-dimensional finite element numerical model. The model was developed by using computation codes set up by ENEL-CRIS. Twenty node isoparametric finite elements were used for a total amount of 1192 nodes. The mesh adopted is shown in the overall view on Fig.8.

Figure 8 Mesh adopted in the three-dimensional finite element mathematical model

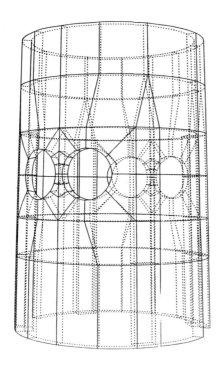

The following characteristics were used for the material:

$E = 30,000$ MPa

$\nu = 0.16$

$\gamma = 0.025$ N/cm^3.

Load conditions were as those of the physical model.

Results were in good agreement with those obtained by the physical model, as it appears in the following table, which compares the results of the two models at more significant points.

4. CONCLUSIONS

What described above is an example in which the adoption of model analysis as a design tool proved to be essential and resolving for the designers. The careful analysis of stresses between hole and hole and around the openings, along with a good agreement of the results achieved

Comparison of stresses resulting at El. 1.555
from the two model analysis

Stresses (MPa)	Dead load		Wind parallel hole axis		Wind perpendicular hole axis	
	phys	math	phys	math	phys	math
Between the holes	-12.8	-11.6	- 2.4 -23.2	- 1.6 -25.1	-14.0	-12.2
Above the gates	- 1.1	- 1.6	- 1.4	- 1.6	+ 0.2 - 2.4	+ 0.6 - 2.8

by the two models, supported the particular design choices, represented by the need of carrying out a special inlet and outlet system for flues.

31 Shear reinforcement in flat plate roofs

A. ABDEL-RAHMAN, M. KHATER and M. NASSEF
Structural Engineering Department, Cairo University, Egypt

SUMMARY

The objectives of this investigation are to study the efficiency of different types of shear reinforcement in flat plate roofs, and to assess the simplicity of fixing such reinforcement to the normal reinforcement required for flexure. The effects of the shear reinforcement on the cracking load, crack width, ultimate load and ductility are also examined.

Five test slabs were prepared. The models simulate the region of negative bending moment around an internal column supporting a flat-plate roof. Different possible methods of shear reinforcement were implemented in the various test models. The test loads were applied incrementally till the ultimate carrying capacity was reached. Deformations, crack widths, crack patterns and ultimate loads were recorded. The test results indicated conservatism of the shear capacity of slab-column connections as specified by the ACI 318 - 77 code and by the British Standards CP110.

Finite element idealization of the test model was utilised to obtain the internal moments and the deformations. Different methods of column load application were considered. The comparison between the theoretical and the experimental values checks the validity of the different methods of column - load application to the theoretical model.

INTRODUCTION

A series of investigations had been initiated since 1954 to examine the strengthening of flat plate column connection by shear reinforcement. In 1976, tests were conducted by (1) on a series of eight half-scale models of reinforced concrete interior flat-plate column specimens. It was found that the use of bent bars as shear reinforcement resulted in an increase in the strength of the connection, but did not cause an increase in the ductility. However, the use of closed stirrups around the straight bar reinforcement that pass through the column, improved the strength and the ductility of the connection. (2) suggested shear reinforcement consisting of short segments (1.2 to 2.5 cms) cut from structural steel I. - beam. This shear reinforcement can be easily placed with little interference with the flexural reinforcement. Six full scale models of slab-column connection were tested. The slabs were simply supported along the edges; an axial load was applied on the central column. This type of shear reinforcement gave a sufficient increase in the ductility and strength of the connection.

In 1980, (3) investigated three different types of preassempled shear reinforcement of the following shapes:

(a) I-segments obtained by cutting short slices from standard I-beam,
(b) Headed shear studs and
(c) Welded wire fabric.

They constructed seven slab - column test specimens which were simply supported and axially loaded. For test specimens utilizing the first two types of shear reinforcement, the failure crack on the compression face took place outside the zone of the shear reinforcement. Welded wire fabric with double unit lead to the same pattern of failure. When single units only were used, the failure crack occurred at the column face. Hence, it was concluded that the shear reinforcements of insufficient anchorage in the tension zone provide no significant increase to shear resistance.

In 1982 (4) carried out experimental investigation using different types of shear reinforcement amongst which were closed and triangular stirrups.

Significant increase in the shear resistance and ductility of the slab-column connection was gained.

The previously tested slabs included expensive types of shear reinforcement. The closed and open stirrups offer less expensive and simpler type of shear reinforcement. The present investigation is aimed at studying the effect of these types. Different shapes are suggested and examined with regard to strength and ease of construction.

EXPERIMENTAL INVESTIGATION

Test Slabs

Five slabs simply supported along the four sides and axially loaded through a central column were prepared. The dimensions are shown in Fig.1. The specimen represents the region of negative bending moment around an interior column and the simply supported edges simulate the lines of contraflexure. The flexural reinforcement consisted of mild steel (37) plane bars of 12 mm diameter in the tension side and 10 mm diameter in the compression side as shown in Fig.2. The five slabs were divided into groups. Slab S_1 did not contain shear reinforcement. It was considered a control specimen. Slabs S_2 and S_3 contained closed and open stirrups respectively. The stirrups were of 6 mm diameter bars enclosing the top and bottom flexural reinforcement of the slab to ensure the effective anchorage as shown in Fig.3. and Fig.4. Slabs S_4 and S_5 included open and triangular stirrups respectively. The stirrups were of 6 mm diameter bar enclosing the tension reinforcement only and embedded in the compression zone as shown in Fig.5. and Fig.6. The concrete used for the preparation of the previous models consisted of locally available gravel, sand, ordinary portland cement and water. The mix was designed to give cube compressive strength of 35 MPa.

Loading Set Up And Measuring Devices

The tested slabs were mounted as shown in Fig.1. The simple support along each edge was provided by steel rods of 22 mm diameter resting on a loading frame, made of structural steel I-Beams No.16. All slabs were axially loaded through the central column by a Hydraulic Jack of 130 tons capacity. Demec Mechanical strain gauges of 20 cm gauge length with accuracy of 0.002 m were used to measure the strains on the bottom and top surfaces of the slab. Deflections on the bottom surface of the slab were measured by dial gauges of an accuracy of 1/100 mm. An optical ultra lens with an accuracy of 1/20 mm was used to measure the crack width.

EXPERIMENTAL AND THEORETICAL RESULTS

The relationship between the applied load and central deflection of the specimen slab S_1 which did not contain shear reinforcement is shown **Fig.7**. The load-deflection relationship takes the form of straight line up till initial cracking load, and deviates from linearity from the initial cracking load till the punching failure load.

The load-deflection curves for test slabs S_2, S_3, S_4 and S_5 are similar as shown in Fig.8. Four stages are distinguished on the typical load-deflection curve; Stage I. starts from the beginning of loading up till the first crack. Stage II. starts from the initial cracking load up till the first yielding. In this stage, flexural cracking of the slab develops. Stage III. starts after the first yielding and up till the ultimate flexural strength as calculated by yield line theory; in this stage yielding of the tension reinforcement spreads from the loaded area towards the slab edges. Stage IV. it is the final stage in which a plastic state of rapidly increasing deflections at no additional load occurs.

Considering that the central deflection at ultimate load is a measure of the ductility and referring to the load deflection relationships shown in Fig.7., it is found that slab S_1 exhibits limited ductility and slabs S_2, S_3, S_4 and S_5 which contained shear reinforcement possess higher ductility.

The strain distribution for slab S_1 as shown in Fig.9.a shows that the tension reinforcement in the vicinity of the column yields before punching; and the failure occurred before the outer bars had not reached their yield stress.

Inspection of the recorded strain distributions of Fig.9. indicates that the maximum strains occurred at the column corners. This confirms the experimental results of (5) in which he noted that the applied concentrated load on the slab was largely transferred from the column to the slab at the column corners.

The load strain diagram for all the tested slabs show that the strain is proportional to the load up till initial cracking of the slab. At about 50% of ultimate load an abrupt change occurred in the rate of increase of the strain in the tension steel. The load at that change is considered the inclined-cracking load.

The initial flexural cracks recorded for all the tested slabs were in the vicinity of the column stub along the lines of the tension reinforcement. These cracks were much more pronounced along lines parallel to the reinforcement close to the tension surface. At about 40% of the ultimate load of slab S_1, the recorded surface strains at the column corners were high enough to cause initial yielding at the level of reinforcement. At about half the ultimate load the slab corners lifted up. At about 65% of the ultimate load the maximum crack width reached 0.3 mm, as may be noticed from Table 1. This load is considered the service-ability limit load for the tested slabs.

The failure for slab S_1 was punching shear failure attained at a load of 16 tons. For slabs S_2, S_3, S_4 and S_5 the cracks + 1 ton = 10 KN. radiated from the vicinity of the column towards the slab edges as the load was incrementally increased. The crack patterns are shown in Fig.10. The radial cracks had reached the slab edges, and the slab corners lifted up at a load of about 50% of the failure load. The final failure of slab S2 was recognised as flexural failure at a load of 24.5 tons. The corresponding maximum crack width was 0.19 cm. and maximum central deflection was 3.4 cms. Disintegration of the compression surface of slabs S_3, S_4 and S_5 had taken place in the vicinity of the column at loads of 23.5, 23. and 22.5 tons respectively. This mode of failure is known as shear-compression failure.

The computed values of shear strength for the model slab S_1 (without shear reinforcement) by the ACI 318 - 77 code and the British code CP110 are 11.2 tons and 10.0 tons respectively. The two estimates are less than the experimental value by 30% and 37% respectively. The upper limit of the nominal shear strength specified by the ACI code (16.8 tons) for slabs containing shear reinforcements was exceeded by at least 34% in the model slabs S_2 to S_5.

Finite element study has been carried out to analyse the tested slabs. Only one quarter of the model was investigated because of symmetry. It was divided into 121 rectangular plate bending elements as shown Fig.11. Five different methods of modelling the load transfer from the column to the slab were investigated as shown in Fig.11. Each load case was studied at a load of 5 tons which is just below the initial cracking load of slab S_1. The results were not significantly affected by the method of distributing the column load.

CONCLUSIONS

The following conclusions may be drawn from the present investigation:

1. The slab-column connection without shear reinforcement had little ductility. Failure occurred suddenly by diagonal tension, cracking and splitting of the concrete along the bars in the tension side of the slab.

2. The predicted values of shear strength for slabs with or without shear reinforcements by the ACI 318 - 77 code and by the British Standards CP110 are conservative.

3. Vertical stirrups enclosing the top and bottom longitudinal reinforcement of the slab are effective in increasing the shear strength and the ductility of slab-column connections, stirrups of this type complicate the placement of reinforcement.

4. Effective anchorage for shear reinforcement consisting of stirrups can be obtained even without tying the stirrups around the flexural steel in the compression side of the slab, this simplifies the placement of shear reinforcement in flat plates by insertion from the top after all the flexural reinforcement are placed.

5. The initial cracking load and the service load corresponding to a maximum crack width of 0.3 mm increased by using stirrups as shear reinforcement.

Table (1)

Experimental Results

Model Slab	The max ultimate load (ton)+	Initial cracking load (ton)	Load at max. crack width = 0.3 mm (ton)	Deflection at max. Load. (cm)
S_1 (without shear reinfor.)	16	5.	9.5	1.35
S_2 (with shear reinfor.)	24.5	6.5	13.25	3.4
S_3	23.5	5.5	13.25	2.85
S_4	23	5.5	13	2.95
S_5	22.5	6.5	13.5	2.75

+ 1 ton ≈ 10 KN

FINAL REMARKS

The deflection and moment values obtained from finite element analysis using the different methods of modelling the column load are more or less the same. The simplest case of modelling for loading is that using point loads at the nodal points corresponding to the four corners of the column.

REFERENCES

(1) ISLAM, S, and PARK, R, 'Tests on slab-column connections with shear and unbalances flexure', Journal of St. Div., ASCE, (March 1976).

(2) LANGOHR, P, and GHALI, A, 'Special shear reinforcement for concrete flat plates', ACI Journal, (March 1976).

(3) SEIBLE, F, and GHALI, A, 'Preassembled Shear Reinforcement units for flat plates', ACI Journal, (Jon., 1980).

(4) PILLAI, U, KIRK, W, and SCARUZZO, L, 'Shear reinforcement at slab-column connections in a r.c. flat plat structures', ACI Journal, (Jan. 1982).

(5) MOE, J, 'Shearing strength of reinforced concrete slabs and footings under concentrated load', PCA, Bulletin D 47, skokie, Ill., U.S.A. (April 1961).

(6) 'Building code requirements for reinforced concrete', ACI-318 - 77, Detroit, ACI, (1977).

(7) 'Code of Practice for the structural use of concrete', CP110, British Standards Institution, (Nov. 1972).

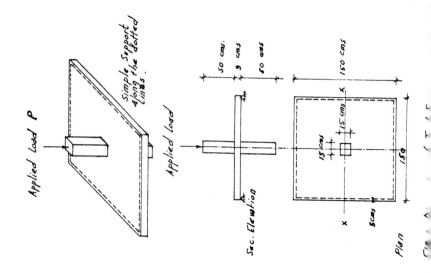
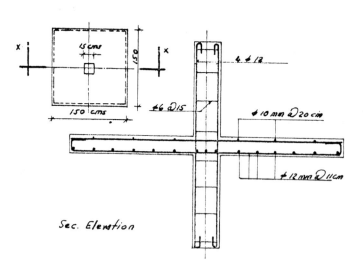

FIG 2. Details of Flexural Reinforcements

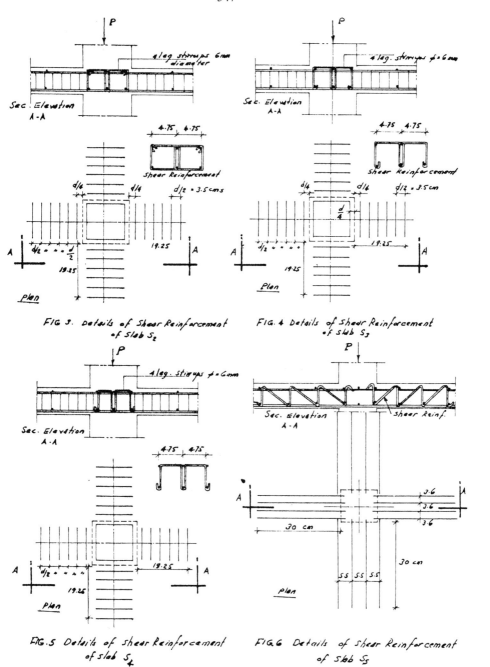

FIG 3. Details of Shear Reinforcement of slab S_2

FIG. 4 Details of Shear Reinforcement of slab S_3

FIG. 5 Details of Shear Reinforcement of slab S_4

FIG. 6 Details of Shear Reinforcement of slab S_5

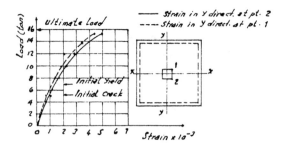

FIG 7 Load Versus Maximum Strain - Slab S_1

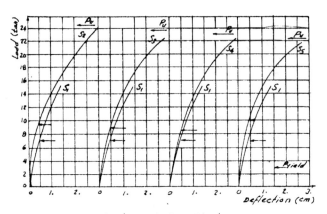

FIG. 8 Load - Central Deflection Diagrams

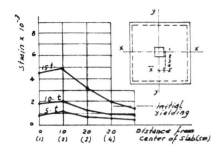

FIG. 9.a Strain Distribution For Slab S_1 along Line $x = \bar{x}$

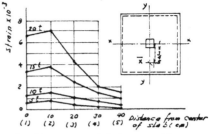

FIG. 9.b Strain Distribution For Slab S_2 along Line $x = \bar{x}$

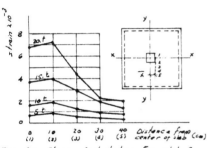

FIG. 9.c Strain Distribution For Slab S_3 along Line $x = \bar{x}$

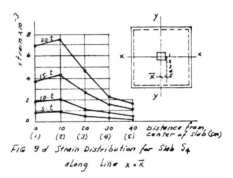

FIG. 9.d Strain Distribution For Slab S_4 along Line $x = \bar{x}$

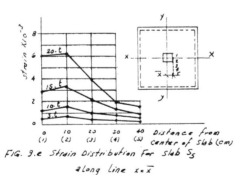

FIG. 9.e Strain Distribution For Slab S_5 along Line $x = \bar{x}$

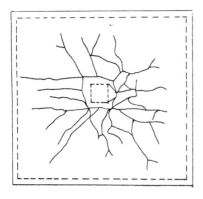

FIG. 10a Cracking Pattern of Slab S_1 at 90% of Ultimate Load

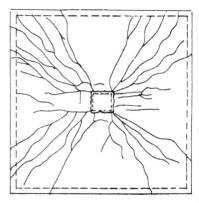

FIG 10b Cracking Pattern of Slab S_2 at 94% of Ultimate Load

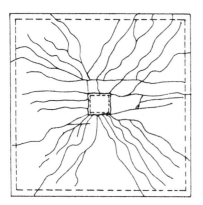

FIG 10·c Cracking Pattern of Slab S_3 at 94% of Ultimate Load

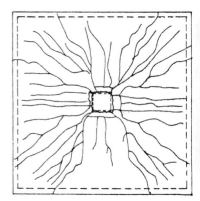

FIG 10·d Cracking Pattern of Slab S_4 at 95% of Ultimate Load

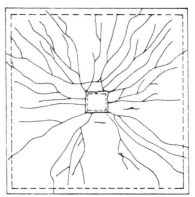

FIG 10·e Cracking Pattern of Slab S_5 at 95%

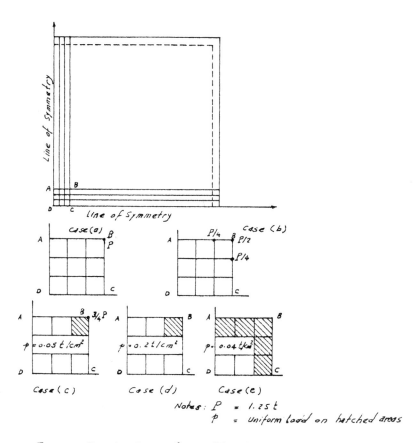

FIG. 11 Idealization of load Transfer Between Column and Slab

DISCUSSION

Dr D J Cleland   University of Belfast   NI

The authors distinguish the failures as a pure punching shear for Slab S_1, flexural for Slab S_2 and shear-compression for Slabs S_3, S_4 and S_5. On what basis were these distinctions made? For example it would be useful to know if shape or size of the "punched cone" differed in each case.

Further the authors state that at 50% of ultimate (ie about 11 kN) in Slabs S_2, S_3, S_4 and S_5 the radial cracks had extended to meet the slab edges whereas Fig 10a indicates that at 14.4 kN in Slab S_1 the cracking was not so excessive. Would the authors wish to comment?

Author's Reply

The failure of Slab S1 was accompanied by sudden displacement of the shear cone relative to the remainder of the slab. This failure was, therefore, characterised as punching failure. In the case of Slab S2, the stiffness deteriorated progressively as the load approached the ultimate capacity. The central defection was 34 mm and the crack width was 19 mm. The failure was then distinguished as flexural failure. Disintegration of the compression surface of Slabs S3, S4 and S5 had taken place under loads below the flexural capacity. The failure was characterised as shear compression failure for these slabs.

The cracking in Slab S1 did not cover a large area due to the formation of wide cracks along the perimeter of the punched cone. In comparison to other slabs, these cracks were localised but much wider.

32 Static test of model RC beams in hydrostatic pressure surrounding

N. K. SUBEDI and D. C. BELL
University of Dundee, UK

SUMMARY

As more and more concrete structures are taking shape for use in the oceans the understanding of the behaviour of reinforced concrete elements under such environment is becoming urgent. Ten identical model beams were tested in flexure; five in hydrostatic pressure surrounding and five in normal atmospheric conditions. The results from the preliminary test and its implications in design are discussed.

INTRODUCTION

The use of reinforced concrete in offshore structures date from about the period of the Second World War[1] when marine structures were used as forts for defence purposes. In shallow water construction the main problems in design appear to be the durability of concrete exposed to marine environment and the corrosion of reinforcement. The effect of hydrostatic pressure is very small.

Now it is conceivable[2] that the future generation of offshore structures will be built in deeper and more hostile waters. The hydrostatic pressure can no longer be assumed negligible. The limit state concept of design requires the knowledge of the behaviour of structural elements from the inception of loading to final failure. For serviceability requirements the prediction and control of cracks and excessive deformations and the effects of hydrostatic pressure upon them are of prime importance. Therefore the basic question to be asked is do the simple structural elements behave the same way in hydrostatic pressure surrounding as they do in the normal atmospheric conditions? At present the answer to this question is far from clear. Most of the design guides dealing with concrete offshore structures are still at the developmental stage.[3,4] It would be some time before these can be adopted as codes of practice. Little else has been published so far in the important field of structural behaviour under hydrostatic environment.

With this background in mind a research programme was started at Dundee University to study the behaviour of basic structural elements in pressurised hydrostatic surrounding. In this paper the results of the test on ten r.c. model beams of identical construction are discussed. Five of these beams were tested in a hydrostatic pressure tank at 0.483 N/mm^2 (70 psi) and the other five in normal atmospheric conditions. All the beams in this series were tested in flexure. The two groups of beams were compared with respect to their ultimate load using CP110 and the crack pattern at failure.

PRESSURE TANK

A pressure tank (Fig 1a) was designed to test the model beams in hydrostatic surrounding. The tank was assembled using standard channel sections and steel plates sealed together by gaskets and a chemical sealant. Six viewing panels consisting of 25 mm thick toughened glass plates were incorporated to observe the test in progress. The provision for the removal of one of the end plates facilitated sufficient access for handling. For the loading of the beams a plunger was provided through the top plate. The plunger was connected to a hydraulic jack and a load cell mounted on to a frame (Fig 1b). The hydrostatic pressure in the tank was applied using a proving hand pump connected to the tank through an end plate.

MODEL BEAMS

Ten model beams of the same design were cast for this particular series of test. The overall dimensions of the beam were 100 mm x 50 mm x 1000 mm long giving a span of 900 mm with third point loads at 300 mm apart. The main reinforcement consisted of 2 Nos 6 mm diameter bars (characteristic strength, f_y = 336 N/mm2). In the compression zone 2 Nos 2.64 mm diameter bars (characteristic strength, f_y = 177 N/mm^2) were provided whose primary function was to support the vertical stirrups.

The beams were designed to fail in flexure; details are shown in Fig 2. A nominal mix of concrete to give approximately 45 N/mm^2 strength at 28 days with 6 mm maximum size of aggregates was used. The concrete strengths are given in Table 1.

Table 1 Concrete Strength

Cube No/Beam No	B1	B2	B3	B4	B5	B6	B7	B8	B9	B10
100 mm size										
f_{cu} 28-day N/mm^2	55.0	40.5	41.2	49.0	48.8	54.0	54.5	53.0	49.0	50.5
f_{cu} day of test N/mm^2	65.0	48.6	51.5	66.5	59.0	63.5	64.0	66.7	61.5	60.5

TEST WORK

Pressurised test - Although the pressure tank was designed for operating up to 1 N/mm^2 (145 psi) it was decided to limit the pressure to 0.483 N/mm^2 (70 psi) for this series of test. Prior to test the beams were marked up into grids so that any cracks appearing could be transferred on to the paper in the appropriate position. The test consisted of two parts (a) the 24-hour immersion of the beam at 0.345 N/mm^2 (50 psi) before testing (b) the actual test to failure at a hydrostatic pressure surrounding of 0.483 N/mm^2 (70 psi).

In the final test in general a stepwise increment of 0.5 kN of load was applied. At the end of each increment the beam was inspected and the progress of cracks was monitored. Approximate deflection of the beam at the loading points was recorded by measuring the movement of the loading ram (Fig 1b). The procedure was followed until the beam reached its ultimate load capacity at failure which was deemed to be the maximum load that the beam could withstand.

Dry test - Test in normal atmospheric conditions was carried out under identical procedures on an Avery-Dennison hydraulic press with a capacity of 150 kN. The dial gauges were set in this case on the load distributing beam (Fig 3).

DISCUSSION OF RESULTS

Load deflection characteristics - This is shown in Fig 4. The curve for the dry test indicates a gradual increase in deflection as the load increases progressively. This contrasts slightly with the curve for pressurised test in which the trend of deflection is almost bi-linear.

The cracking load, P_c, was calculated using the Engineer's Formula,

$$P_c = 6(f_t + p) I_e / \left[1(h - h_n)\right] \qquad \qquad \qquad (1)$$

in which, f_t is the uniaxial tensile strength of concrete (assumed 0.1 f_{cu}), p is the hydrostatic pressure, and I_e is the equivalent second moment of area for the uncracked section. Other parameters are defined in Fig 4c.

It would appear from Fig 4a and 4b that after the initial cracking, load-deflection characteristics for the two types of test are similar.

Ultimate load and crack pattern - Except in one case (beam B3) the mode of failure of the beams was as expected, flexural with the main cracks in the middle third of the span. In beam B3 diagonal tension was the cause of failure. From the crack patterns there seems very little to differentiate between the pressurised test and the dry test. The beams after failure are compared in Fig 5.

The theoretical ultimate load and the test results are compared in columns 4 to 6, Table 2. The calculated values were based on the compatibility of strain and a rectangular stress diagram with a maximum stress of 0.67 f_{cu}. In general (column 6) the beams tested in the normal atmospheric conditions have a bigger margin of safety than those tested in the pressurised hydrostatic environment.

Table 2 Comparison of Ultimate Load

Beam	1 Test Dry or Pressurised	2 Position of N-axis, x	3 Mon. of Resistance	4 Ultimate Load W_{calc}	5 W_{test}	6 W_{calc}/W_{test}
	D or P	mm	Nmm x 10^6	N	N	Ratio
B1	P	9.50	1.572	10479	10000	1.04
B2	D	11.50	1.564	10423	10665	0.98
B3	P	11.00	1.550	10333	10400	0.99
B4	D	9.25	1.576	10505	13156	0.80
B5	P	10.00	1.564	10427	10200	1.02
B6	D	9.50	1.572	10479	12566	0.85
B7	P	9.50	1.572	10479	10800	0.97
B8	D	9.25	1.576	10505	12658	0.83
B9	P	9.75	1.568	10452	10600	0.99
B10	D	9.75	1.568	10452	12160	0.86

	Average ratio	W_{calc}/W_{test}
	P →	1.00
	D →	0.86

The average ratio for the pressurised test is 1.00 compared to 0.86 for the ... indicates that the ultimate strength of beams exposed to pr... ironment is roughly 14% less than that of beams in normal at... nditions. It must however be judged as a preliminary finding ... llowing limitations (a) the number of beams tested constituted ... ll number (b) the hydrostatic pressure in the tank was only 0 ... (70 psi). This is a very small pressure representing approximat... pth of water (c) the test was a short term test.

CONCLUSION

1. From ... liminary test carried out on model beams it suggests that the ultimate load carrying capacity of flexural beam elements in a hydrostatic pressure environment is slightly lower than in the normal atmospheric conditions.

2. Further tests at higher pressures are required before making any recommendations for the design of prototype structural elements.

ACKNOWLEDGEMENT

The tests described here were carried out as part of the Honours year project of the second author in the Department of Civil Engineering, Dundee University. The Authors acknowledge the support given by the technical staff in the Civil Engineering Laboratory throughout the test work.

REFERENCES

(1) 'Concrete in the oceans - A research programme into the use of concrete in the North Sea', CIRIA/UEG, C & CA, Dept of Energy.

(2) CLARKE, J L, 'Concrete offshore structures - the next ten years', Concrete, March 1983.

(3) BSI 'Draft for development - fixed offshore structures', DD55: 1978.

(4) DNV 'Rules for the design, construction and inspection of offshore Structures', 1977.

(5) BELL, D C, 'Static loading of r.c. beams in a pressurised hydrostatic environment', Honours Report, Dept of Civil Engineering, University of Dundee, April 1984.

(a) Pressure tank with beam specimen after the test

Figure 1   Pressurised test

(b) Loading arrangement

Figure 3   Model beam under test in normal atmospheric conditions

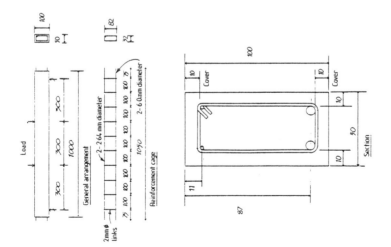
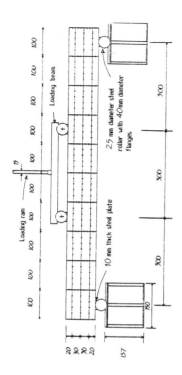

Figure 2 Details of beams

(a)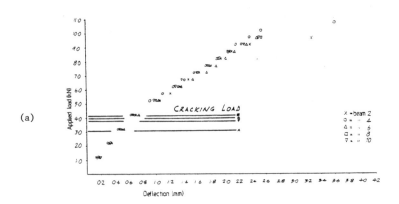

(b)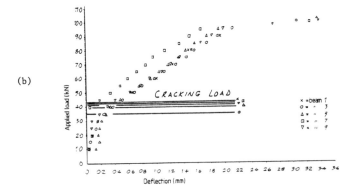

(c)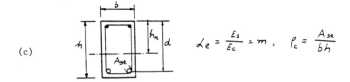

Figure 4 Load-deflection characteristics
(a) normal test   (b) pressured test
(c) uncracked section

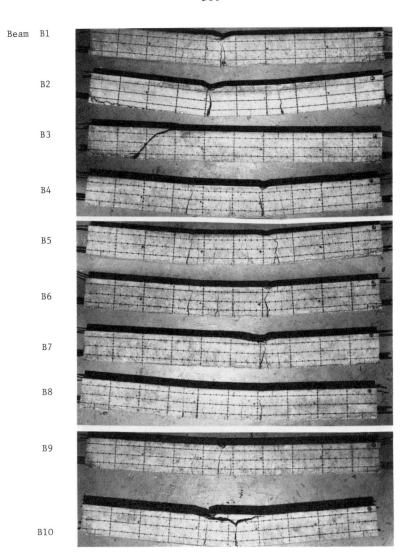

Figure 5 Model beams after failure: beams B1, B3, B5, B7 and B9 subjected to pressurised test; beams B2, B4, B6, B8 and B10 subjected to normal test.

33 Comparative analysis of concrete perforated tubes monolithically cast and made by precasting

MIRCEA MIHAILESCU, IOAN OLARIU, VIORICA BUDIU and
NICOLAE POIENAR
Polytechnic Institute Cluj-Napoca, Romania

SUMMARY

The main purpose of the experimental investigations carried out on two perforated-tube models, the first cast in place and the second one made by precasting, was to clarify some aspects concerning the behavior of this types of structures, subjected to lateral statical and dynamical actions.

INTRODUCTION

Insofar as the lateral load resisting function of the buildings is concerned, three broad types of units may be distinguished: frames, walls and tubes. The first two types of units made the construction of taller buildings relatively expensive and/or technically inadequate, due to theirs insufficient stiffness to resist lateral loads. The natural tendency then was to find new systems that would utilize the perimeter configurations of such buildings (1). The development of the framed-tube system was, therefore, a logical outcome of this intention.
The framed-tube involves two distinct types of structural behavior as follows: (i)frame behavior of the two walls parallel to the direction of the horizontal load, and (ii)tube behavior of the entire structure (2). The greatest advantage of the framed-tube system is that it conforms to the traditional architectural arrangement of the windows and its use may be economically and

aesthetically justified for a wide range of number of stories.
The main purpose of this paper was to clarify some aspects
concerning the behavior of the framed-tube structure (cast in
place or made by precasting), subjected to lateral loads.
Computer calculations and laboratory tests under static and
dynamic loads were made.

MODELS CHARACTERISTICS AND INVESTIGATION PROGRAM

Models characteristics

The two models, the first cast in place (3),(4), and the second
one made of two type of segments assembled by posttensioning (5),
represent the 1:30- scale reduction of a 20-story frame-tube
prototype structure. The dimensions of the rectangular sections
were 49,42x66 cm. (Fig.1) and the height was 200 cm.
The 1,5 cm. thick walls had four vertical rows of 20 openings
(5,3 x 7,3 cm.) on the short side and six such vertical rows of
20 openings on the long side. Hollowed slabs 1,2 cm. thick served
as diaphragms at every story along the structure's height.
The controlling parameters of the frame-tube, as defined by Khan
and Amin (6) are: stiffness ratio= 0,503, stiffness factor= 0,306
and aspect ratio= 0,673.

Investigation program

The investigation program intended to specify:
- the magnitude of real bending and torsion stiffness of the
framed-tube.
- the settlement of the bending model as well as the equivalent
"channel flanges" necessary for preliminary design.
- the determination of the dynamical characteristics of the
structure subjected to bending and torsion oscilations (only for
the cast in place model).
The experimental results were compared with those obtained by the
static and dynamic analysis, performed by the ETABS computer
program (8), adapted for use in the FELIX C-256- type computer (7)

MODELS EXECUTION AND TESTING TECHNIQUES

Models execution

Microconcrete with maximum gravel size 2mm., 600 kg. ordinary portland cement per cubic meter and 0,45 water/cement ratio was used to manufacture the models. The mechanical characteristics of the microconcrete are: 10 cm. cube strength- 32,5 N/mm^2, prism strength - 27,5 N/mm^2, tensile strength - 0,3 N/mm^2, modulus of elasticity - 2,2 x 10^4 MN/m^2.

The main reinforcement of the current columns consisted from 6$\emptyset$2,5 mm. mild wires and of the corner columns 12$\emptyset$2,0 mm. vertical wires, while the horizontal reinforcement of the spandrels consisted from 6$\emptyset$2,0 mm. wires. The columns and spandrels were also provided with closed rectangular stirrups of 1,0 mm. diameter mild steel wires at a constant spacing of 10,0 mm. The yield strength of the reinforcement was 220 N/mm^2.

The monolithically poured model

Due to the small dimensions, the execution of the model was performed by cutting it in two halves with a symmetric vertical plane, following three main steps:
- the precasting of the slabs;
- the cast in situ of the two halves of the model embeding in the walls the edges of the precast slabs;
- the achievement of the connections between the two parts by splicing the reinforcement and casting in situ the points (Fig.2)

The model made by precasting

The formworth and the reinforcements of the two types of precast units are presented in Fig.3. Fig.4 shows the field segments after form stripping.

Special attention was paid to the performance of the joints:
- the vertical joints were made by splicing the reinforcement and casting in situ the points.

- the horizontal joints were strengthed by the shear thresholds provided at each element and by the SBP ∅ 2,5 mm. prestressing wires, anchored in the foundation and crossing through all the precast units of the 20 stories (Fig. 5). The posttensioning force was 100 daN in each wire.

Testing techniques

The triangular steel skeletons provided with pulley wheels were symmetricaly placed along the loading axis of the model.
The horizontal forces (in three loading steps: 0,7; 1,12 and 1,5 kN) were created by means of dead weights suspended on strings passing over the pulley wheels and having the horizontal branche ends connected to the model at the desired floor.
The torsional loading was provided by two equal, oposite and parallel point loads in the plane containing the desired floor.
The lateral deflections and rotations were measured by means of 1/100 mm. reading accuracy dial gauges fixed on a sustaining steel skeleton arround the model. The strain distribution in two horizontal sections of the wall was recorded with resistance strain gauges conected to a strain bridge.
The bending and torsion-free oscillations were achieved by releasing the initial forced horizontal deflections and rotations induced at the top of the model. An oscillograph SM311 coupled with a videocorder 8LS-1 was used to record the accelerations response of the models, detected by KD16 capacitive transducers. The forced oscillations were achieved with a 4 daN sinusoidal excitator coupled to an impulse generator (1-100 Hz).

PRESENTATIONS OF RESULTS AND CONCLUSIONS

Lateral loading

Typical deflection graphs for the 70 daN loading applied at the top of the models are shown in Fig. 6 for the cast in place model and for the precast model. There are plotted also the deflections calculated considering various dimensions of the equivalent "channel flanges". It is obvious that, due to the posttensioning, the stiffness of the model made by precasting is aprox. 30 % greather than the stiffness of the cast in place model. For the preliminary design of the cast in places model it is necessary to consider I2 equivalent channel f̄   ᵐʰʸle for the precast framed-tube an I3 section is suit

Torsion loading

The torsion rotations plotted in Fi̟   w a good agreement between the experimental and theoret   For the preliminary design it is possible toreplace the   )e by an equivalent orthotropic tube whose cross-sect   quals the sum of the columnar cross-sectional areas   dulus beeng determined by the Coull and Bose relatio   ;ations evaluated for this equivalent tube represe   ation of about 15 % as compared to the real model

Oscillations

The bending and torsion fre̠   ̠s at 20 and 15 levels are presented in Fig. 8. The fundamental p̠̠ iod of bending oscillations may be computed for the preliminary design considering the same channel flanges as for the lateral loadings. The response at the forced vibration, plotted in Fig. 9, confirms this result.
A relatively good approximation for the torsion period of oscillations can be abtained using the equivalent orthotropic tube (as for torsion loading), considering a one mass system, lumped at H/2 high.

REFERENCES

(1) KHAN, F R, "Analysis and design of framed tube structures for tall concrete buildings", SP-36, ACI, Detroit, (1973).

(2) ACI COMMITTEE 442, "Response of buildings to lateral forces", ACI Journal, (1972).

(3) OLARIU, I, Interaction problems in tall building analysis, Doctoral thesis, Politechnical Institute, Cluj-Napoca, (1980).

(4) MIHAILESCU, M, OLARIU, I, POCANSCHI, A, "Model investigation of a twenty stories frame-tube structure", *Proceedings of the second national symposium of experimental stress analysis* Cluj-Napoca, (1980).

(5) BUDIU, N B, *Structuri prefabricate în formă de tub perforat pentru clădiri etajate*, Doctoral thesis, Politechnical Institute, Cluj-Napoca, (1984).

(6) KHAN, F R, AMIN, NAVINCHANDRA, R, "Analysis and design of framed tubes structures for tall buildings", *The structural engineer*, Vol. 51, No. 3, (Mar. 1973).

(7) OLARIU, I, "ETABS- a three dimensional analysis frame-shearwall system computer program", *Proceedings of the first symposium of applied programming*, Sibiu, Vol. 2, (1978).

(8) WILSON, E L, HOLLINGS, J P, DOVEY, H H, "Three dimensional analysis of building systems", *Report EERC 75-13*, Berkeley, (1975).

(9) POCANSCHI, A, OLARIU, I, Discussion of "Orthotropic membrane for tall building analysis", *Proceedings ASCE*, Vol. 105, (Oct. 1979).

(10) POCANSCHI, A, OLARIU, I, "Response of a medium-rise frame-tube model under static and dynamic actions", *ACI Journal*, (March-April 1982).

(11) COULL, A, BOSE, B, "Simplified analysis of framed-tube structures", *Proceedings ASCE*, Vol. 1o1, (Nov. 1975).

Figure 1   Geometric characteristics.

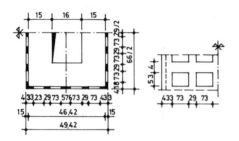

Figure 2   Cast in place model.

Figure 3   Formwork and reinforcement of the precast units.

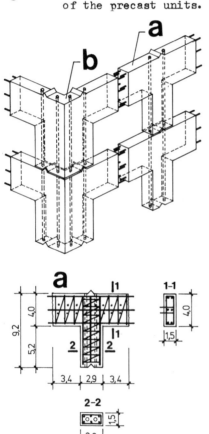

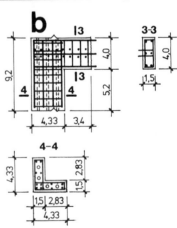

Figure 4 Precast units

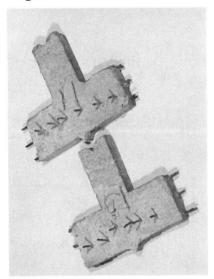

Figure 5 Precast model

Figure 6 Bending deflections

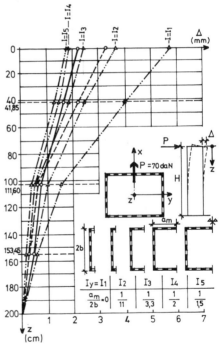

Figure 7 Torsion rotations   Figure 8 Free oscillations

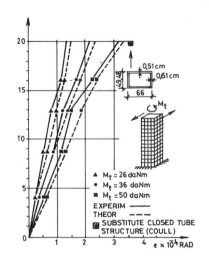
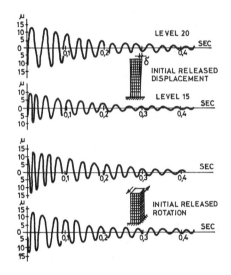

Figure 9 Forced vibrations response

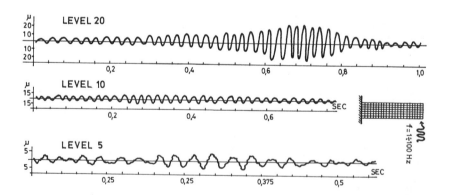

Section 4 Papers 24–33

Commentary by RICHARD N. WHITE
(Professor of Structural Engineering, Cornell University, Ithaca, New York, USA)

This section, the final one in our two-day meeting on use of models in design, contained an excellent set of interesting papers on a diverse set of topics. The reported results show in a convincing manner that models do have a very important role in modern structural engineering. Rather than comment on each of the individual papers, it may be more useful to provide a perspective on physical modelling, focussing on the problems met in the modelling of concrete structures and also offering some comments on where the future applications of structural modelling may lie.

Given that the goal of using models is to improve design methodology, which in turn leads to better constructed facilities with improved reliability, economy and serviceability, we should turn to models whenever they can make a positive and effective contribution to the state-of-the-art of structural engineering.

The classic subdivision of modelling into Elastic and Inelastic models, and into Static and Dynamic Models, provides a convenient starting point for these brief comments. The role of each type in modern structural engineering, and in the future, will be addressed subsequently. Models can be further classified into Design Models, Research Models and Instructional Models. Design models are by definition intended to provide answers to specific questions, while research models lead to better codes and to confirmation of improved analytical capabilities; all of this occurs because the physical modelling provides us with new insight into complex behaviour that cannot be predicted well by analysis alone. Instructional models are used too little in modern education, by incorporating their usage into structural engineering coursework, and by having laboratory work in physical modelling of structures, we can produce young engineers with a stronger physical sense of structural performance. It is my opinion that this type of education will be critical to the engineer of the future who will be executing most of his or her design work with the computer.

Models can be looked at from the category of prototype structure they are based upon. Models of Conventional Structures are best suited to consider local behaviour of design details and to understand the three-dimensional behaviour of the complete structural system, which includes potential instability modes. Models of Unconventional Structures (complex geometries and complex loading environments) serve a somewhat different purpose in that they are necessary to advance our basic understanding of the behaviour of the structure as it responds to load. Offshore structures and nuclear power structures often fall into this latter category.

What are the critical problems met in modelling concrete structures? In spite of identification of many of these problems several decades ago, we are still striving to come up with improved techniques. As shown by the papers at this seminar, we have not yet reached final consensus on the best method for scaling prototype concrete. Scaling of aggregate, methods of best scaling tensile, and size effects continue to be topics of meaningful research in the area of model concretes. The use of fracture mechanics concepts has great potential in providing some definitive answers to the puzzle of size effect in shear strength of reinforced and prestressed concrete, and it is recommended that research efforts in this area be accelerated. Once the shear size effect is fully understood, we still will face the problem of how to handle it in reduced scale models. But we will at least be in a much better position to properly define the use of models in shear-sensitive situations.

Bond behaviour of reinforcing steel is finally nearing some state of reasonable understanding. The work reported in this seminar, along with similar work at the Kajima Institute of Construction Technology in Japan, shows that ribbed models bars can in fact properly simulate full-scale reinforcing. The availability of this type of deformed model reinforcing in a variety of sizes and at several strength levels would give substantial impetus to the physical modelling community; a persistent problem has been how we might be able to pool our resources to underwrite the production and storage of quantities of the reinforcement.

Problems in loading and instrumentation tend to focus on two areas - dead load similitude requirements and dynamic loading. Self-weight effects must be modelled in many situations and are particularly difficult to achieve in soil mechanics problems and in massive structures; continued development of centrifuge facilities will lead to important advances. Similarly, the building of earthquake simulators has enabled many model studies to be made of concrete structures under seismic loadings. The range of applicability of small scale models to shake-table loading environments is still being established. Given that we will never be able to do as many large-scale tests as we need to fully understand dynamic response of concrete structures, reduced scale tests will be essential, and the loading devices and associated instrumentation must be at the highest level.

Potential lack of acceptance of model test results by other professionals cannot be ignored. This problem can be countered from several avenues: (a) by additional comprehensive critical comparisons of results of tests on different size "identical" structures, (b) by clarification of the cause of size effects, and (c) by our taking realistic positions on what models can offer and what their limitations are.

Future applications of physical models for concrete structures will certainly include many studies of the three-dimensional nature of structural systems, and the role of the various load-carrying mechanisms as loading progresses and inelastic behaviour becomes important. The entire range of building and bridge forms needs studies of this type, probably mostly in the research mode where the associated results can be generalised and fed back into the design process. This will call for modelling of the very highest level.

Earthquake response of above ground structures and of buried facilities (such as life-lines) will utilise models to an increasing degree, with loadings being both dynamic and pseudo-dynamic. Dynamic and quasi-static response of off-shore and coastal structures to wave forces and ice loading has received considerable attention in recent years, and again reduced-scale models can be expected to pay important roles in studying these complex three-dimensional interaction problems.

Wind effects on structures has received little mention at this seminar because the techniques are not peculiar to the material of construction. This area of modelling will continue to place strong reliance on the elastic model.

New generations of nuclear power plant structures will also demand sophisticated physical models. Much of the technology for gas-cooled reactors was based on model tests done in the United Kingdom, France and elsewhere, and structural problems in future designs will likewise need extensive testing to verify analytical approaches and to study behaviour.

While other structural types come to mind, this brief discussion will end here with the comment that there is no substitute for physical understanding of structural behaviour. Seminars such as the one we are closing here today are critical elements in advancing the state-of-the-art of structural engineering and to furthering our quest for the "perfect structure" that can be built for minimum cost and will perform to the owner's satisfaction over its entire lifetime.

List of delegates

NAME	REPRESENTING
A Abdel-Rahman A	Cairo University Egypt
D P Abrams D P	University of Colorado USA
Dr T J A Agar	Mott Hay and Anderson UK
Dr W F Anderson	University of Sheffield UK
G S T Armer	Building Research Establishment UK
P Barr	UK Atomic Energy Authority UK
Dr L F Boswell	The City University UK
Miss S H Buchner	University of Surrey UK
P Chana	Cement & Concrete Association UK
H K Cheong	Imperial College UK
Dr J L Clarke	Cement & Concrete Association UK
Dr D J Cleland	Queen's University Belfast NI
M J G Connell	Property Services Agency UK
Dr R J Cope	University of Liverpool UK
Dr W Gene Corley	Portland Cement Association USA
R J Craig	New Jersey Institute of Technology USA
J P Cripps	British Telecom UK
Dr C D'Mello	The City University UK
Dr M Dostal	National Nuclear Corporation Ltd UK
Prof Dr-Ing J Eibl	University of Karlsruhe W Germany
B R Ellis	Building Research Establishment UK
Dr H M Emam	Cairo University Egypt
Dr J Fairbairn	National Engineering Laboratory UK
F M Falih	University of Sheffield UK
Dr C C Fleischer	Taylor Woodrow Construction Ltd UK
Prof E Fumagalli	ISMES Italy
Dr F K Garas	Taylor Woodrow Construction Ltd UK
Prof J E Gibson	The City University UK
Dr C D Goode	University of Manchester UK
J Herter	Bundesanstalt fur Materialprufung (BAM) W Germany
W D Howe	UK Atomic Energy Authority
R Iliya	Dar-Al-Handasah Consultants Egypt
Dr Imperato	ISMES Italy
Dr R Jiminez	University of Pureto Rico Puerto Rico
J Jowett	UK Atomic Energy Authority UK
T Krauthammer	University of Minnesota USA
J P Leeson	British Telecom UK

Dr S Y A Ma	Mott Hay and Anderson UK
S Majlessi	North East London Polytechnic UK
Dr J B Menzies	Building Research Establishment UK
R J W Milne	Institute of Structural Engineers UK
P D Moncarz	Failure Analysis Associates USA
P L T Morgan	National Nuclear Corporation Ltd UK
Prof E H Morsy	Research Centre for Housing Building & Planning, Cairo Egypt
Prof Dr-Ing R K Muller	University of Stuttgart W Germany
Prof Dr M M El-A Nassef	Cairo University Egypt
A J Neilson	UK Atomic Energy Authority UK
J Nielsen	Technical University of Denmark Denmark
Dr F A Noor	North East London Polytechnic UK
Dr S H Perry	Imperial College UK
D M Porter	University College Cardiff UK
S Raveendran	North East London Polytechnic UK
Dr A G Reid	University of Strathclyde UK
H Rezai-Jorabi	Polytechnic of Central London UK
C M Romer	Property Services Agency UK
Dr G Somerville	Cement & Concrete Association UK
Dr R N Swamy	University of Sheffield UK
W Thoma Dipl-Ing	University of Stuttgart W Germany
J J A Tolloczko	Wimpey Offshore Engineers & Constructors Ltd UK
Prof D N Trikka	University of Roorkee India
Prof Dr-Ing H Twelmeier	Institut fur Statik -TU Braunschweig W Germany
Dr P Waldron	University of Bristol UK
Dr A J Watson	University of Sheffield UK
Dr R G H Watson	Building Research Establishment UK
R N White	Cornell University USA
Dr C Williams	Plymouth Polytechnic UK
Dr J G M Wood	Mott Hay and Anderson UK
H D Wright	University College Cardiff UK

Index of contributors

Abdel-Rahman, A., 338
Abrams, D.P., 138
Al-Azawi, T.K., 158
Al-Shaikh, A.H., 59
Ambraseys, N.N., 59
Anderson, W.F., 268

Bell, D.C., 353
Boswell, L.F., 277
Brandes, K., 168
Brandmann, D., 208
Budiu, V., 362

Carvalhal, F.J., 79
Carvalho, E.C., 311
Castoldi, A., 329
Chana, P., 105
Cheong, H.K., 59
Cleland, D.J., 192
Cope, R.J., 257
Corley, W.G., 129
Craig, R.J., 12
Curzon, A.M., 247

D'Mello, C., 277

Eibl, J., 230
Evans, H.R., 38
Evans, W., 69

Falih, F.M., 25
Feyerabend, E.M., 230

Gibson, J.E., 92
Gilbert, S.G., 192
Giuseppetti, G., 329
Gogate, A.B., 220

Harding, P.W., 38
Herter, J., 168
Hiraishi, H., 129

Jervis Pereira, J.M., 79
Jimenez-Perez, R., 220

Khater, M., 338

Kim, W., 181
Krauthammer, T., 296
Krawinkler, H., 300

Limberger, E., 168
Loy, O., 12

Majlessi, S., 117
Mihailescu, M., 362
Moncarz, P.D., 247
Morgan, B.J., 129
Müller, R.K., 1, 149
Munch-Andersen, J., 320

Nassef, M., 338
Newman, J.B., 117
Nielsen, J., 320
Noor, F.A., 69, 117

Olariu, I., 362
Osteraas, J.D., 247

Poienar, N., 362
Pompeu Santos, S., 311

Raveendran, S., 69
Ruggeri, L., 329
Ryden, C., 12

Sabnis, G., 220
Subedi, N.K., 353
Sulaiman, J.A., 268
Swamy, R.N., 25

Thoma, W., 149
Twelmeier, H., 208

Uliana, F., 329

Wallace, B., 300
Watson, A.J., 158
White, R.N., 48, 181
Wright, H.D., 38

Yong, K.Y., 268

Subject index

Actions
 bending, 16, 25, 120–1, 208, 363
 bond, 7–8, 139, 141
 buckling, 53, 208 ff.
 compression, 15, 283
 dowel, 218–19
 hydrostatic, 353
 hysteresis, 126–7, 129, 138
 impact, 149 ff., 158, 168
 punching shear, 25, 35, 220
 shear, 16, 201–2, 222, 257, 294, 338
 tension, 15, 283
 torsion, 181, 363

Bond
 effects, 180
 tests, 7–8, 147

Costs
 scale models, of, 250–1

Instrumentation
 displacement transducers, 213–14, 281
 electrical resistance strain gauges, 48, 52, 281, 333

Loading
 cyclic, 59, 313
 ground vermiculite, 50
 impact, 149 ff., 158, 168, 230
 oyster shell, 324
 powdered rock, 50
 precooked rice, 50
 prestressing, 193, 364

Loading—*contd.*
 rape seed, 324
 rice, 50
 seismic, 79, 137, 140, 300 ff., 314
 transient, 59 ff., 131

Materials
 concrete, 14, 120, 128
 control tests, 7, 10–11, 13, 40, 118–20, 183
 fly ash, 26, 33
 ground vermiculite, 50
 gypsum, 12, 126, 183
 microconcrete, 4–6, 10–11, 159, 183, 331
 plastic sheeting, 48
 powdered rock, 50
 precooked rice, 50
 steel, 6–7, 27, 139, 159, 174, 331

Microconcrete
 additives, 11
 aggregate, 10
 fly ash, 26, 33
 glass bend, 73
 mix design, 4–6, 26–7, 119, 153, 179, 183, 210, 271, 331, 364
 strength, 5, 10–11
 Young's modulus, 5, 150

Model structures
 arches, 290 ff.
 beam/column joints, 126–7, 138, 301
 beams, 12 ff., 72, 117 ff., 126, 153, 168, 208, 230, 353

Model structures—*contd.*
  bin, 48 ff.
  box, 29 ff.
  chimney, 329 ff.
  columns, 12, 230
  composite beams, 38
  frames, 138, 317
  grouted connections, 277
  pile cap, 220
  piles
    bored, 268
    cast *in situ*, 268
  RC containment, 247 ff.
  silo, 48 ff., 320 ff.
  slab/column connections, 25, 92 ff., 192 ff., 311, 338
  slabs, 156, 168, 173, 181, 192, 230, 338
  T-beams, 155–7
  wall assemblies, 129, 300
Modelling problems
  aggregate interlock, 218
  cracking behaviour, 30, 111, 160, 193–5, 252
  curing, 107
  fabrication, 107
  flow in silos, 320
  size effects, 92, 110, 114–16, 117 ff., 126, 128 ff., 138

Modelling problems—*contd.*
  strain
    gradient, 107
    rate, 147, 174

Reinforcement
  bond tests, 7–8
  deformed bars, 69 ff., 151
  fibre, 12 ff., 24
  high yield, 69
  knurling, 28, 147, 150, 210
  scaling law, 2–4
  steel wire, 6–7, 27, 139, 159, 174, 331
  welded struts, 39 ff., 182–4, 190

Scaling
  bin model, 44–55, 57
  concrete structure, 105 ff.
  model loading material, 50
  modelling laws, 2–4, 49–50, 149
  silo model, 49–50, 57
  slab/column connections, 25 ff., 96–8
  slabs, 164–5, 181

Size effects
  model tests, in, 92, 110, 114–16, 117 ff., 126, 128, 138